T0276051

Lecture Notes in Mathematics

2151

More information about this series at http://www.springer.com/series/304

Saint-Flour Probability Summer School

The Saint-Flour volumes are reflections of the courses given at the Saint-Flour Probability Summer School. Founded in 1971, this school is organised every year by the Laboratoire de Mathématiques (CNRS and Université Blaise Pascal, Clermont-Ferrand, France). It is intended for PhD students, teachers and researchers who are interested in probability theory, statistics, and in their applications.

The duration of each school is 13 days (it was 17 days up to 2005), and up to 70 participants can attend it. The aim is to provide, in three high-level courses, a comprehensive study of some fields in probability theory or Statistics. The lecturers are chosen by an international scientific board. The participants themselves also have the opportunity to give short lectures about their research work.

Participants are lodged and work in the same building, a former seminary built in the 18th century in the city of Saint-Flour, at an altitude of 900 m. The pleasant surroundings facilitate scientific discussion and exchange.

The Saint-Flour Probability Summer School is supported by:
– Université Blaise Pascal
– Centre National de la Recherche Scientifique (C.N.R.S.)
– Ministère délégué à l'Enseignement supérieur et à la Recherche

For more information, see
http://recherche.math.univ-bpclermont.fr/stflour/stflour-en.php

Christophe Bahadoran
bahadora@math.univ-bpclermont.fr

Arnaud Guillin
Arnaud.Guillin@math.univ-bpclermont.fr

Laurent Serlet
Laurent.Serlet@math.univ-bpclermont.fr

Université Blaise Pascal – Aubière cedex, France

Zhan Shi

Branching Random Walks

École d'Été de Probabilités
de Saint-Flour XLII – 2012

Zhan Shi
Laboratoire de Probabilités et Modèles
  Aléatoires
Université Pierre et Marie Curie
Paris, France

ISSN 0075-8434   ISSN 1617-9692   (electronic)
Lecture Notes in Mathematics
ISBN 978-3-319-25371-8   ISBN 978-3-319-25372-5   (eBook)
DOI 10.1007/978-3-319-25372-5

Library of Congress Control Number: 2015958655

Mathematics Subject Classification: 60J80, 60J85, 60G50, 60K37

Springer Cham Heidelberg New York Dordrecht London

Printed on acid-free paper

Springer International Publishing AG Switzerland is part of Springer Science+Business Media
(www.springer.com)

To the memory of my teacher,

Professor Marc Yor (1949–2014)

但去莫复问，白云无尽时。
－ [唐] 王维《送别》

Preface

These notes attempt to provide an elementary introduction to the one-dimensional discrete-time branching random walk and to exploit its spinal structure.

They begin with the case of the Galton–Watson tree for which the spinal structure, formulated in the form of the size-biased tree, is simple and intuitive.

Chapter 3 is devoted to a few fundamental martingales associated with the branching random walk.

The spinal decomposition is introduced in Chap. 4, first in its more general form, followed by two important examples. This chapter gives the most important mathematical tool of the notes.

Chapter 5 forms, together with Chap. 4, the main part of the text. Exploiting the spinal decomposition theorem, we study various asymptotic properties of the extremal positions in the branching random walk and of the fundamental martingales.

The last part of the notes presents a brief account of results for a few related and more complicated models.

The lecture notes by Berestycki [43] and Zeitouni [235] give a general and excellent account of, respectively, branching Brownian motion and the F-KPP equation and branching random walks with applications to Gaussian free fields.

I would like to deeply thank Yueyun Hu; together we wrote about 20 papers in the last 20 years, some of them strongly related to the material presented here. I am grateful to Élie Aïdékon, Julien Berestycki, Éric Brunet, Xinxin Chen, Bernard Derrida, Gabriel Faraud, Nina Gantert, and Jean-Baptiste Gouéré for stimulating discussions, to Bastien Mallein and Michel Pain for great assistance in the preparation of the present notes, and to Christian Houdré for correcting my English with patience.

I wish to thank Laurent Serlet and the Scientific Board of the École d'été de probabilités de Saint-Flour for the invitation to deliver these lectures.

Paris, France   Zhan Shi
August 2015

Contents

Chapter 1
Introduction

We introduce branching Brownian motion as well as the branching random walk, and present the elementary but very useful tool of the many-to-one formula. As a first application of the many-to-one formula, we deduce the asymptotic velocity of the leftmost position in the branching random walk. The chapter ends with some examples of branching random walks and more general hierarchical fields.

1.1 Branching Brownian Motion

Branching Brownian motion is a simple continuous-time spatial branching process defined as follows. At time $t = 0$, a single particle starts at the origin, and moves as a standard one-dimensional Brownian motion, whose lifetime, random, has the exponential distribution of parameter 1. When the particle dies, it produces two new particles (in other words, the original particle splits into two), moving as independent Brownian motions, each having a mean 1 exponential random lifetime. The particles are subject to the same splitting rule. And the system goes on indefinitely. See Fig. 1.1.

Let $X_1(t), X_2(t), \ldots, X_{N(t)}(t)$ denote the positions of the particles in the system at time t. Let

$$f(x) := \mathbf{1}_{\{x \geq 0\}}.$$

We consider

$$u(t, x) := \mathbf{E}\left(\prod_{i=1}^{N(t)} f(x + X_i(t)) \right).$$

© Springer International Publishing Switzerland 2015
Z. Shi, *Branching Random Walks*, Lecture Notes in Mathematics 2151,
DOI 10.1007/978-3-319-25372-5_1

Fig. 1.1 Branching
Brownian motion

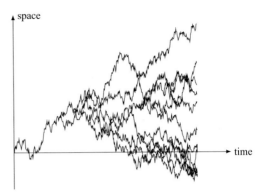

By conditioning on the lifetime of the initial ancestor, it is seen that

$$u(t, x) = e^{-t} \mathbf{E}[f(x + B(t))] + \int_0^t e^{-s} \mathbf{E}[u^2(t - s, x + B(s))] \, ds$$

$$= e^{-t} \mathbf{E}[f(x + B(t))] + e^{-t} \int_0^t e^r \mathbf{E}[u^2(r, x + B(t - r))] \, dr, \quad (r := t - s)$$

where $(B(s), s \geq 0)$ denotes standard Brownian motion. We then arrive at the
so-called F-KPP equation (Fisher [113] who was interested in the evolution of a
biological population, Kolmogorov et al. [158])

$$\frac{\partial u}{\partial t} = \frac{1}{2} \frac{\partial^2 u}{\partial x^2} + u^2 - u. \tag{1.1}$$

This equation holds for a large class of measurable functions f. The special form
of f we take here is of particular interest, since in this case,

$$u(t, x) = \mathbf{P}\left(\min_{1 \leq i \leq N(t)} X_i(t) \geq -x \right) = \mathbf{P}\left(\max_{1 \leq i \leq N(t)} X_i(t) \leq x \right),$$

which is the distribution function of the maximal position of branching Brownian
motion at time t.

The F-KPP equation is known for its travelling wave solutions: let $m(t)$ denote
the median of u, i.e., $u(t, m(t)) = \frac{1}{2}$, then

$$\lim_{t \to \infty} u(t, x + m(t)) = w(x),$$

uniformly in $x \in \mathbb{R}$, and w is a wave solution of the F-KPP equation (1.1) at speed
$2^{1/2}$, meaning that $w(x - 2^{1/2}t)$ solves (1.1), or, equivalently,

$$\frac{1}{2} w'' + 2^{1/2} w' + w^2 - w = 0.$$

It is proved by Kolmogorov et al. [158] that $\lim_{t \to \infty} \frac{m(t)}{t} = 2^{1/2}$, and by Bramson [67, 69] that

$$m(t) = 2^{1/2}t - \frac{3}{2^{3/2}}\ln t + C + o(1), \quad t \to \infty, \tag{1.2}$$

for some constant C.

There is a probabilistic interpretation of the travelling wave solution w: by Lalley and Sellke [164], w can be written as

$$w(x) = \mathbf{E}\left(e^{-C_1 D_\infty e^{-2^{1/2}x}}\right), \tag{1.3}$$

where $C_1 > 0$ is a constant, and $D_\infty > 0$ is a random variable whose distribution depends on the branching mechanism (in our description, it is binary branching). The idea of this interpretation is also present in the work of McKean [180].

The connection, observed by McKean [180], between the branching system and the F-KPP differential equation makes the study of branching Brownian motion particularly appealing.[1] As such, branching Brownian motion can be used to obtain—or explain—results for the F-KPP equation. For purely probabilistic approaches to the study of travelling wave solutions to the F-KPP equation, see Neveu [204], Harris [123], Kyprianou [162]. More recently, physicists have been much interested in the effect of noise on wave propagation; see discussions in Sect. 6.2.

We study branching Brownian motion as a purely probabilistic object. Moreover, the Gaussian displacement of particles in the system does not play any essential role, which leads us to study the more general model of branching random walks.

1.2 Branching Random Walks

These notes are devoted to the (discrete-time, one-dimensional) branching random walk, which is a natural extension of the Galton–Watson process in the spatial sense. The distribution of the branching random walk is governed by a random N-tuple $\Xi := (\xi_i, 1 \le i \le N)$ of real numbers, where N is also random and can be 0; alternatively, Ξ can be viewed as a finite point process on $\mathbb{R}$.

An initial ancestor is located at the origin. Its children, who form the first generation, are scattered in $\mathbb{R}$ according to the distribution of the point process Ξ. Each of the particles (also called individuals) in the first generation produces its own children who are thus in the second generation and are positioned (with respect to their parent) according to the same distribution of Ξ. The system goes on

---

[1]Another historical reference is a series of papers by Ikeda et al. [141–143], who are interested in a general theory connecting probability with differential equations.

Fig. 1.2 A branching
random walk and its first four
generations

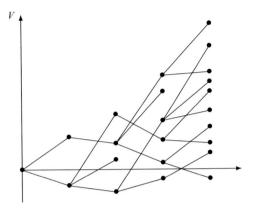

indefinitely, but can possibly die if there is no particle at a generation. As usual, we
assume that each individual in the n-th generation reproduces independently of each
other and of everything else until the n-th generation. The resulting system is called
a **branching random walk**. See Fig. 1.2.

We mention that several particles can share a same position.

It is clear that if we only count the number of individuals in each generation, we
get a Galton–Watson process, with $N := \#\Xi$ governing its reproduction distribution.

Throughout, $|x|$ denotes the generation of the particle x, and x_i (for $0 \le i \le |x|$)
denotes the ancestor of x in the i-th generation (in particular, $x_0 := \varnothing$, $x_{|x|} := x$).
Let $(V(x), |x| = n)$ denote the positions of the individuals in the n-th generation.
We are interested in the asymptotic behaviour of $\inf_{x:\,|x|=n} V(x)$.

Let us introduce the (log-)Laplace transform of the point process[2]

$$\psi(t) := \ln \mathbf{E}\left(\sum_{x:\,|x|=1} e^{-tV(x)} \right) \in (-\infty, \infty], \quad t \in \mathbb{R}. \tag{1.4}$$

We always assume that $\psi(0) > 0$, and that $\psi(t) < \infty$ for some $t > 0$.

The assumption $\psi(0) > 0$, i.e., $\mathbf{E}(\#\Xi) > 1$, means that the associated Galton–
Watson tree is supercritical, so by Theorem 2.1 in Sect. 2.1, the system survives with
positive probability. However, $\psi(0)$ is not necessarily finite.

The assumption $\inf_{t>0} \psi(t) < \infty$ ensures that the leftmost particle has a linear
asymptotic velocity, as we will see in Theorem 1.3 in Sect. 1.4.

---

[2]For notational simplification, we often write from now on $\inf_{|x|=n}(\cdots)$ or $\sum_{|x|=1}(\cdots)$, instead
of $\inf_{x:\,|x|=n}(\cdots)$ or $\sum_{x:\,|x|=1}(\cdots)$, with $\inf_{\varnothing}(\cdots) := \infty$ and $\sum_{\varnothing}(\cdots) := 0$.

1.3 The Many-to-One Formula

Throughout this section, we fix $t > 0$ such that $\psi(t) < \infty$.

Let $S_0 := 0$ and let $(S_n - S_{n-1}, \ n \geq 1)$ be a sequence of independent and identically distributed (i.i.d.) real-valued random variables such that for any measurable function $h : \mathbb{R} \to [0, \infty)$,

$$\mathbf{E}[h(S_1)] = \frac{\mathbf{E}[\sum_{|x|=1} e^{-tV(x)} h(V(x))]}{\mathbf{E}[\sum_{|x|=1} e^{-tV(x)}]},$$

i.e., $\mathbf{E}[h(S_1)] = \frac{\mathbf{E}[\sum_{u \in \Xi} e^{-tu} h(u)]}{\mathbf{E}[\sum_{u \in \Xi} e^{-tu}]}$ if you prefer a formulation in terms of the point process Ξ.

Theorem 1.1 (The Many-to-One Formula) *Assume that $t > 0$ is such that $\psi(t) < \infty$. For any $n \geq 1$ and any measurable function $g : \mathbb{R}^n \to [0, \infty)$, we have*

$$\mathbf{E}\left[\sum_{|x|=n} g(V(x_1), \ldots, V(x_n)) \right] = \mathbf{E}\left[e^{tS_n + n\psi(t)} g(S_1, \ldots, S_n) \right].$$

Proof We prove it by induction in n. For $n = 1$, this is the definition of the distribution of S_1. Assume the identity proved for n. Then, for $n + 1$, we condition on the branching random walk in the first generation; by the branching property, this yields

$$\mathbf{E}\left[\sum_{|x|=n+1} g(V(x_1), \ldots, V(x_{n+1})) \right]$$

$$= (\mathbf{E} \otimes \widetilde{\mathbf{E}})\left[\sum_{|y|=1} \sum_{|\tilde{z}|=n} g(V(y), V(y) + \widetilde{V}(\tilde{z}_1), \ldots, V(y) + \widetilde{V}(\tilde{z}_n)) \right],$$

where $\widetilde{\mathbf{E}}$ is expectation with respect to the branching random walk $(\widetilde{V}(\tilde{z}))$ which is independent of $(V(y), |y| = 1)$. By induction hypothesis, for any $u \in \mathbb{R}$,

$$\widetilde{\mathbf{E}}\left(\sum_{|\tilde{z}|=n} g(u + \widetilde{V}(\tilde{z}_1), \ldots, u + \widetilde{V}(\tilde{z}_n)) \right) = \widetilde{\mathbf{E}}\left(e^{t\widetilde{S}_n + n\psi(t)} g(u, u + \widetilde{S}_1, \ldots, u + \widetilde{S}_n) \right),$$

with the random walk $(\widetilde{S}_j, \ j \geq 1)$ independent of $(V(y), |y| = 1)$, and distributed as $(S_j, \ j \geq 1)$ under $\mathbf{P}$. Since

$$\mathbf{E}\left[\sum_{|y|=1} h(V(y)) \right] = \mathbf{E}\left[e^{tS_1 + \psi(t)} h(S_1) \right],$$

it remains to note that $(\mathbf{E} \otimes \widetilde{\mathbf{E}})[e^{tS_1 + t\widetilde{S}_n + (n+1)\psi(t)} g(S_1, S_1 + \widetilde{S}_1, \ldots, S_1 + \widetilde{S}_n)]$ is nothing else but $\mathbf{E}[e^{tS_{n+1} + (n+1)\psi(t)} g(S_1, S_2, \ldots, S_{n+1})]$. This implies the desired identity for all $n \geq 1$. □

Remark 1.2 Behind the innocent-looking new random walk (S_n) is a change-of-probabilities setting, which we will study in depth in Chap. 4. □

1.4 Application: Velocity of the Leftmost Position

Now that we are equipped with the many-to-one formula, let us see how useful it can be via a simple application. As we will prove deeper results in the forthcoming chapters, our concern here is not to provide arguments in their full generality. Rather, we focus, at this stage, on understanding of how the many-to-one formula can help us in the study of the branching random walk.

Among the most immediate questions about the branching random walk, is whether there is an asymptotic velocity of the extreme positions. The answer is as follows.

Theorem 1.3 *Assume $\psi(0) > 0$. If $\psi(t) < \infty$ for some $t > 0$, then almost surely on the set of non-extinction,*

$$\lim_{n \to \infty} \frac{1}{n} \inf_{|x|=n} V(x) = \gamma, \tag{1.5}$$

where

$$\gamma := - \inf_{s>0} \frac{\psi(s)}{s} \in \mathbb{R}. \tag{1.6}$$

Remark 1.4 If instead we want to know the velocity of $\sup_{|x|=n} V(x)$, we only need to replace the point process Ξ by $-\Xi$. □

Before proving Theorem 1.3, we need a simple lemma, stated as follows.

Lemma 1.5 *For any $k \geq 1$ and $t > 0$, we have*

$$\frac{1}{k} \mathbf{E}\left[\inf_{|x|=k} V(x) \right] \geq - \frac{\psi(t)}{t}.$$

Proof We have

$$\frac{1}{k} \mathbf{E}\left[- \inf_{|x|=k} t V(x) \right] \leq \frac{1}{k} \ln \mathbf{E}\left[e^{- \inf_{|x|=k} t V(x)} \right] \quad \text{(Jensen's inequality)}$$

$$\leq \frac{1}{k} \ln \mathbf{E}\left[\sum_{|x|=k} e^{-t V(x)} \right]. \quad \text{(bounding max by sum)}$$

It remains to note that $\mathbf{E}[\sum_{|x|=k} e^{-tV(x)}] = e^{k\psi(t)}$ by the many-to-one lemma or by a direct computation. $\qquad\square$

It is now time to prove Theorem 1.3.

Proof of Theorem 1.3 We prove the theorem under the additional condition that $\#\varXi \geq 1$ a.s. (i.e., the system survives with probability one) and that $\psi(0) < \infty$.

Let $X_n := \inf_{|x|=n} V(x)$. It is easily seen that for any pair of positive integers n and k, we have

$$X_{n+k} \leq X_n + \widetilde{X}_k ,$$

where $\widetilde{X}_k$ is a random variable having the distribution of X_k, independent of X_n. This property does not allow us to use Kingman's subadditive ergodic theorem. However, we can use an improved version [166] to deduce that $\frac{X_n}{n} \to \alpha$ a.s. and in L^1, with $\alpha := \inf_{n \geq 1} \frac{\mathbf{E}(X_n)}{n}$.

So we need to check that $\alpha = \gamma$. By Lemma 1.5, we have $\alpha \geq \gamma$; so it remains to check that $\alpha \leq \gamma$. Let $\varepsilon > 0$. It suffices to prove that for some integer $L = L(\varepsilon) \geq 1$ and with positive probability,

$$\limsup_{j \to \infty} \frac{1}{jL} \inf_{|x|=jL} V(x) \leq - \inf_{s>0} \frac{\psi(s)}{s} + \varepsilon .$$

Let $t > 0$ be such that

$$\frac{\psi(t)}{t} > \psi'(t) > \inf_{s>0} \frac{\psi(s)}{s} - \varepsilon.$$

[Let $s^* := \inf\{s > 0 : \frac{\psi(s)}{s} = \inf_{u>0} \frac{\psi(u)}{u}\} > 0$. If $0 < s^* < \infty$, then we only need to take $t \in (0, s^*)$; if $s^* = \infty$, which means that $\inf_{s>0} \frac{\psi(s)}{s}$ is not reached but is the limit when $s \to \infty$, then any $t \in (0, \infty)$ will do.]

Write $a := \inf_{s>0} \frac{\psi(s)}{s} - \varepsilon$. We construct a new Galton–Watson tree $\widetilde{\mathbb{T}}$ which is a subtree of the original Galton–Watson tree $\mathbb{T}$: the first generation of the new Galton–Watson tree $\widetilde{\mathbb{T}}$ consists of all the vertices x in the L-th generation of $\mathbb{T}$ such that $V(x) \leq -aL$; more generally, for any integer $n \geq 1$, if x is a vertex in the n-th generation of $\widetilde{\mathbb{T}}$, its offspring in the $(n+1)$-th generation of $\widetilde{\mathbb{T}}$ consists of all the vertices y in the $(n+1)L$-th generation of $\mathbb{T}$ which are descendants of x in $\mathbb{T}$ such that $V(y) - V(x) \leq -aL$.

By construction, the new Galton–Watson tree $\widetilde{\mathbb{T}}$ has mean offspring $m_{\widetilde{\mathbb{T}}} := \mathbf{E}[\sum_{|x|=L} \mathbf{1}_{\{V(x) \leq -aL\}}]$. By the many-to-one formula,

$$m_{\widetilde{\mathbb{T}}} = \mathbf{E}\left[e^{tS_L + L\psi(t)} \mathbf{1}_{\{S_L \leq -aL\}} \right].$$

Recall that $\psi'(t) < \frac{\psi(t)}{t}$, so let us choose and fix $b \in (\psi'(t), \frac{\psi(t)}{t})$, to see that

$$m_{\widetilde{\mathbb{T}}} \ge e^{[\psi(t)-bt]L} \, \mathbf{P}(-bL \le S_L \le -aL).$$

Since $\mathbf{E}(S_1) = -\psi'(t)$ by our choice of t (which lies in $(0, s^*)$), we have $-b < \mathbf{E}(S_1) < -a$ by definition, so that $\mathbf{P}(-bL \le S_L \le -aL) \to 1, L \to \infty$, whereas $e^{[\psi(t)-bt]L} \to \infty, L \to \infty$. Therefore, we can choose and fix L such that $m_{\widetilde{\mathbb{T}}} > 1$.

The new Galton–Watson tree $\widetilde{\mathbb{T}}$ being supercritical, it has positive probability to survive (by Theorem 2.1 in Sect. 2.1). Upon the set of non-extinction, we have $\inf_{|x|=jL} V(x) \le -ajL = (-\inf_{s>0} \frac{\psi(s)}{s} + \varepsilon)jL, \forall j \ge 1$. This completes the proof of Theorem 1.3. $\qquad\square$

We close this section with a question. Theorem 1.3 tells us that the asymptotic velocity of $\inf_{|x|=n} V(x)$ is determined by the Laplace transform ψ. However, the Laplace transform of a point process does not describe completely the law of the point process. For example, it provides no information of the dependence structure between the components. If $\psi(0) < \infty$, then there exists a real-valued random variable ξ such that

$$\psi(t) - \psi(0) = \ln \mathbf{E}[e^{-t\xi}], \quad t \ge 0.$$

Let us consider the following model: the reproduction law of its associated Galton–Watson process is the law of $\#\Xi$; given the Galton–Watson tree, we assign, on each of the vertices, i.i.d. random variables distributed as ξ. We call the resulting branching random walk $(V_\xi(x))$. According to Theorem 1.3, if $0 < \psi(0) < \infty$ and if $\psi(t) < \infty$ for some $t > 0$, then $\frac{1}{n} \inf_{|x|=n} V(x)$ and $\frac{1}{n} \inf_{|x|=n} V_\xi(x)$ have the same almost sure limit.

Question 1.6 Give an explanation for this identity without using Theorem 1.3.

1.5   Examples

We give here some examples of branching random walks, and more general hierarchical fields. In the literature, the branching random walk bears various names, all leading to equivalent or similar structure. Let us make a short list.

Example 1.7 (Mandelbrot's Multiplicative Cascades) Mandelbrot's multiplicative cascades are introduced by Mandelbrot [193], and studied by Kahane [150] and Peyrière [210], in an attempt at understanding the intermittency phenomenon in Kolmogorov's turbulence theory. It can be formulated, for example, in terms of a stochastically self-similar measure on a compact interval. In fact, the standard Cantor set consists in dividing, at each step, a compact interval into three identical sub-intervals and removing the middle one. Instead of splitting an interval into three identical sub-intervals, we can use a possibly random number of sub-intervals according to a certain finite-dimensional distribution (which is not necessarily supported in a simplex, while the dimension can be random), and the resulting lengths of sub-intervals form an example of Mandelbrot's multiplicative cascade. If we look at the logarithm of the lengths, we have a branching random walk.

Mandelbrot's multiplicative cascades also bear other names, such as random recursive constructions [194]. A key ingredient is to study fixed points of the so-called **smoothing transforms** [20, 21, 103]. For surveys on these topics, see Liu [168], Biggins and Kyprianou [58]. □

Example 1.8 (Gaussian Free Fields and Log-Correlated Gaussian Fields) The two-dimensional discrete Gaussian free field possesses a complicated structure of extreme values, but it turns out to be possible to compare it with that of the branching random walk. By comparison to analogous results for branching random walks, many deep results have been recently established for Gaussian free fields and more general logarithmically correlated Gaussian fields [60, 61, 96, 183]. In parallel, in the continuous-time setting, following Kahane's pioneer work in [151], the study of **Gaussian multiplicative chaos** has witnessed importance recent progress [101, 116, 217].

Via Dynkin's isomorphism theorem, local times of Markov processes are closely connected to (the square of) some Gaussian processes. As such, new lights have been recently shed on the **cover time** of the two-dimensional torus by simple random walk [36, 95]. □

Example 1.9 (Spatial Branching Models in Physics) In [94], Derrida and Spohn introduced **directed polymers on trees**, as a hierarchical extension of **Derrida's Random Energy Model** (REM) for spin glasses. In this setting, the energy of a polymer, being the sum of i.i.d. random variables assigned on each edge of the tree, is exactly a branching random walk with i.i.d. displacements. The continuous-time setting has also been studied in the literature [66].

Directed polymers on trees also provide an interesting example of random environment for random walks. The tree-valued random walk in random environment is an extension of Lyons's biased random walk on trees [171, 172], in the sense that the random walk is randomly biased. Chapter 7 will be devoted to this model.

The F-KPP equation has always enjoyed much popularity in the physics literature. For example, in particle physics, high energy evolution of the quantum chromodynamics (QCD) amplitudes is known to obey the F-KPP equation [201]. In Sect. 6.2, we are going to discuss on branching random walks with selection, in connection with the slowdown phenomenon in the wave propagation of the F-KPP equation studied by physicists. For a substantial review on the physics literature of the F-KPP equation, see van Saarloos [228]. □

1.6   Notes

As mentioned in the preface, the lecture notes of Berestycki [43] and Zeitouni [235] give a general and excellent account of, respectively, branching Brownian motion and the F-KPP equation, and branching random walks with applications to Gaussian free fields.

The many-to-one formula presented in Sect. 1.3 can be very conveniently used in computing the first moment. There is a corresponding formula, called the many-to-few formula, suitable for computing higher-order moments; see Harris and Roberts [126].

Theorem 1.3 in Sect. 1.4 is proved by Hammersley [120] in the context of the first-birth problem in the Bellman–Harris process, by Kingman [156] for the positive branching random walk, and by Biggins [49] for the branching random walk.

We assume throughout that $\psi(t) < \infty$ for some $t > 0$. Without this assumption, the behaviour of the minimal position in the branching random walk has a different nature. See for example the discussions in Gantert [114].

The list of examples in Sect. 1.5 should be very, very long (I am trying to say that the present list is very, very incomplete); see Biggins [54] for a list of references dealing with the branching random walk under other names. Let me add a couple of recent and promising examples: Arguin [26] delivers a series of lectures on work in progress on characteristic polynomials of unitary matrices, and on the Riemann zeta function on the critical line, whereas Aïdékon [11] successfully applies branching random walk techniques to Conformal Loop Ensembles (CLE).

Chapter 2
Galton–Watson Trees

We recall a few elementary properties of supercritical Galton–Watson trees, and introduce the notion of size-biased trees. As an application, we give in Sect. 2.3 the beautiful conceptual proof by Lyons et al. [176] of the Kesten–Stigum theorem for the branching process.

The goal of this brief chapter is to give an *avant-goût* of the spinal decomposition theorem, in the simple setting of the Galton–Watson tree. If you are already familiar with any form of the spinal decomposition theorem, this chapter can be skipped.

2.1 The Extinction Probability

Consider a Galton–Watson process, also referred to as a Bienaymé–Galton–Watson process, with each particle (or: individual) having i children with probability p_i (for $i \geq 0$; $\sum_{j=0}^{\infty} p_j = 1$), starting with one initial ancestor. To avoid trivial discussions, we assume throughout that $p_0 + p_1 < 1$.

Let Z_n denote the number of particles in the n-th generation. By definition, if $Z_n = 0$ for a certain n, then $Z_j = 0$ for all $j \geq n$. We write

$$q := \mathbf{P}\{Z_n = 0 \text{ eventually}\}, \qquad \text{(extinction probability)}$$

$$m := \mathbf{E}(Z_1) = \sum_{i=0}^{\infty} i p_i \in (0, \infty]. \qquad \text{(mean number of offspring of each individual)}$$

Theorem 2.1

(i) *The extinction probability q is the smallest root of the equation $f(s) = s$ for $s \in [0, 1]$, where $f(s) := \sum_{i=0}^{\infty} s^i p_i$, $0^0 := 1$.*

(ii) *In particular, $q = 1$ if $m \leq 1$, and $q < 1$ if $1 < m \leq \infty$.*

© Springer International Publishing Switzerland 2015
Z. Shi, *Branching Random Walks*, Lecture Notes in Mathematics 2151,
DOI 10.1007/978-3-319-25372-5_2

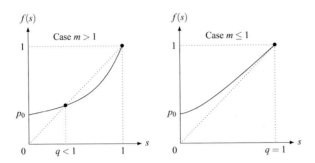

Fig. 2.1 Generating function of the reproduction law

Proof By definition, $f(s) = \mathbf{E}(s^{Z_1})$, and $\mathbf{E}(s^{Z_n} \mid Z_{n-1}) = f(s)^{Z_{n-1}}$. So $\mathbf{E}(s^{Z_n}) = \mathbf{E}(f(s)^{Z_{n-1}})$, which leads to $\mathbf{E}(s^{Z_n}) = f_n(s)$ for any $n \geq 1$, where f_n denotes the n-th fold composition of f. In particular, $\mathbf{P}(Z_n = 0) = f_n(0)$.

Since $\{Z_n = 0\} \subset \{Z_\ell = 0\}$ for all $n \leq \ell$, we have

$$q = \mathbf{P}\Big(\bigcup_n \{Z_n = 0\}\Big) = \lim_{n \to \infty} \mathbf{P}(Z_n = 0) = \lim_{n \to \infty} f_n(0).$$

The function $f : [0, 1] \to \mathbb{R}$ is increasing and strictly convex, with $f(0) = p_0 \geq 0$ and $f(1) = 1$. It has at most two fixed points. Note that $m = f'(1-)$. See Fig. 2.1.

If $m \leq 1$, then $p_0 > 0$, and $f(s) > s$ for all $s \in [0, 1)$. So $f_n(0) \to 1$. In other words, $q = 1$ is the unique root of $f(s) = s$.

If $m \in (1, \infty]$, then $f_n(0)$ converges increasingly to the unique root of $f(s) = s$, $s \in [0, 1)$. In particular, $q < 1$. □

It follows that in the subcritical case (i.e., $m < 1$) and in the critical case ($m = 1$), there is extinction with probability 1, whereas in the supercritical case ($m > 1$), the system survives with positive probability.

If $m < \infty$, we can define

$$M_n := \frac{Z_n}{m^n}, \qquad n \geq 0.$$

Since (M_n) is a non-negative martingale with respect to the natural filtration of (Z_n), we have $M_n \to M_\infty$ a.s., where M_∞ is a non-negative random variable. By Fatou's lemma, $\mathbf{E}(M_\infty) \leq \liminf_{n \to \infty} \mathbf{E}(M_n) = 1$. It is, however, possible that $M_\infty = 0$. So it is important to know whether $\mathbf{P}(M_\infty > 0)$ is positive.

If there is extinction, then trivially $M_\infty = 0$. In particular, by Theorem 2.1, we have $M_\infty = 0$ a.s. if $m \leq 1$. What happens if $m > 1$?

Lemma 2.2 *Assume $m < \infty$. Then $\mathbf{P}(M_\infty = 0)$ is either q or 1.*

Proof We already know that $M_\infty = 0$ a.s. if $m \leq 1$. So let us assume $1 < m < \infty$.

By definition, $Z_{n+1} = \sum_{i=1}^{Z_1} Z_n^{(i)}$ (notation: $\sum_\varnothing := 0$), where $Z_n^{(i)}$, $i \geq 1$, are copies of Z_n, independent of each other and of Z_1. Dividing both sides by m^n and letting $n \to \infty$, it follows that mM_∞ has the law of $\sum_{i=1}^{Z_1} M_\infty^{(i)}$, where $M_\infty^{(i)}$, $i \geq 1$, are copies of M_∞, independent of each other and of Z_1. Hence $\mathbf{P}(M_\infty = 0) = \mathbf{E}[\mathbf{P}(M_\infty = 0)^{Z_1}] = f(\mathbf{P}(M_\infty = 0))$, i.e., $\mathbf{P}(M_\infty = 0)$ is a root of $f(s) = s$, so $\mathbf{P}(M_\infty = 0) = q$ or 1. □

Theorem 2.3 (Kesten and Stigum [155]) *Assume $1 < m < \infty$. Then*

$$\mathbf{E}(M_\infty) = 1 \;\Leftrightarrow\; \mathbf{P}(M_\infty > 0 \,|\, \text{non-extinction}) = 1 \;\Leftrightarrow\; \mathbf{E}(Z_1 \ln_+ Z_1) < \infty,$$

where $\ln_+ x := \ln \max\{x, 1\}$.

Theorem 2.3 says that $\mathbf{E}(M_\infty) = 1 \Leftrightarrow \mathbf{P}(M_\infty = 0) = q \Leftrightarrow \sum_{i=1}^\infty p_i\, i \ln i < \infty$.

The proof of Theorem 2.3 is postponed to Sect. 2.3. We will see that the condition $\mathbf{E}(Z_1 \ln_+ Z_1) < \infty$, apparently technical, is quite natural.

2.2 Size-Biased Galton–Watson Trees

In order to introduce size-biased Galton–Watson trees, let us view the tree as a random element in a probability space $(\Omega, \mathscr{F}, \mathbf{P})$, using the standard formalism.

Let $\mathscr{U} := \{\varnothing\} \cup \bigcup_{k=1}^\infty (\mathbb{N}^*)^k$, where $\mathbb{N}^* := \{1, 2, \dots\}$. For elements u and v of $\mathscr{U}$, let uv be the concatenated element, with $u\varnothing = \varnothing u = u$.

A tree ω is a subset of $\mathscr{U}$ satisfying the following properties: (i) $\varnothing \in \omega$; (ii) if $uj \in \omega$ for some $j \in \mathbb{N}^*$, then $u \in \omega$; (iii) if $u \in \omega$, then $uj \in \omega$ if and only if $1 \leq j \leq N_u(\omega)$ for some non-negative integer $N_u(\omega)$.

In words, $N_u(\omega)$ is the number of children of the vertex u. Vertices of ω are labeled by their line of descent: the vertex $u = i_1 \dots i_n \in \mathscr{U}$ stands for the i_n-th child of the i_{n-1}-th child of $\dots$ of the i_1-th child of the initial ancestor $\varnothing$. See Fig. 2.2.

Fig. 2.2 Vertices of a tree as elements of $\mathscr{U}$

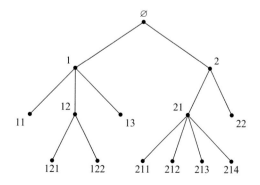

Let Ω be the space of all trees, endowed with a σ-field $\mathscr{F}$ defined as follows. For $u \in \mathscr{U}$, let $\Omega_u := \{\omega \in \Omega : u \in \omega\}$ be the subspace of Ω consisting of all the trees containing u as a vertex. [In particular, $\Omega_\varnothing = \Omega$.] Let $\mathscr{F} := \sigma\{\Omega_u, \ u \in \mathscr{U}\}$.

Let $\mathbb{T} : \Omega \to \Omega$ be the identity application.

Let $(p_k, \ k \geq 0)$ be a probability, i.e., $p_k \geq 0$ for all $k \geq 0$, and $\sum_{k=0}^{\infty} p_k = 1$. There exists a probability $\mathbf{P}$ on $(\Omega, \mathscr{F})$ [203] such that the law of $\mathbb{T}$ under $\mathbf{P}$ is the law of the Galton–Watson tree with reproduction distribution (p_k).

Let $\mathscr{F}_n := \sigma\{\Omega_u, \ u \in \mathscr{U}, \ |u| \leq n\}$, where $|u|$ is the length of u (or the generation of the vertex u in the language of trees). Note that $\mathscr{F}$ is the smallest σ-field containing all the $\mathscr{F}_n$.

For any tree $\omega \in \Omega$, let $Z_n(\omega)$ be the number of individuals in the n-th generation, i.e., $Z_n(\omega) := \#\{u \in \mathscr{U} : u \in \omega, \ |u| = n\}$. It is easily checked that for any n, Z_n is a random variable taking values in $\mathbb{N} := \{0, 1, 2, \ldots\}$.

Assume now $m < \infty$. Since (M_n) is a non-negative martingale, we can define $\mathbf{Q}$ to be the probability on $(\Omega, \mathscr{F})$ such that for any n,

$$\mathbf{Q}_{|\mathscr{F}_n} = M_n \bullet \mathbf{P}_{|\mathscr{F}_n},$$

where $\mathbf{P}_{|\mathscr{F}_n}$ and $\mathbf{Q}_{|\mathscr{F}_n}$ are the restrictions of $\mathbf{P}$ and $\mathbf{Q}$ on $\mathscr{F}_n$, respectively.

For any n, $\mathbf{Q}(Z_n > 0) = \mathbf{E}[1_{\{Z_n>0\}}M_n] = \mathbf{E}[M_n] = 1$, which yields $\mathbf{Q}(Z_n > 0, \ \forall n) = 1$: there is almost sure non-extinction of the Galton–Watson tree $\mathbb{T}$ under the new probability $\mathbf{Q}$. The Galton–Watson tree $\mathbb{T}$ under $\mathbf{Q}$ is called a size-biased Galton–Watson tree. We intend to give a description of its paths.

We start with a lemma. Let $N := N_\varnothing$. If $N \geq 1$, we write $\mathbb{T}_1, \mathbb{T}_2, \ldots, \mathbb{T}_N$ for the N subtrees rooted at each of the N individuals in the first generation.

Lemma 2.4 *Let $k \geq 1$. If $A_1, A_2, \ldots, A_k$ are elements of $\mathscr{F}$, then*

$$\mathbf{Q}(N = k, \ \mathbb{T}_1 \in A_1, \ldots, \mathbb{T}_k \in A_k)$$

$$= \frac{kp_k}{m} \frac{1}{k} \sum_{i=1}^{k} \mathbf{P}(A_1) \cdots \mathbf{P}(A_{i-1})\mathbf{Q}(A_i)\mathbf{P}(A_{i+1}) \cdots \mathbf{P}(A_k). \qquad (2.1)$$

Proof By the monotone class theorem, we may assume, without loss of generality, that $A_1, A_2, \ldots, A_k$ are elements of $\mathscr{F}_n$, for some n. Write $\mathbf{Q}_{(2.1)}$ for $\mathbf{Q}(N = k, \ \mathbb{T}_1 \in A_1, \ldots, \mathbb{T}_k \in A_k)$. Then

$$\mathbf{Q}_{(2.1)} = \mathbf{E}\Big(\frac{Z_{n+1}}{m^{n+1}} 1_{\{N=k, \ \mathbb{T}_1 \in A_1, \ldots, \mathbb{T}_k \in A_k\}}\Big).$$

On $\{N = k\}$, we can write $Z_{n+1} = \sum_{i=1}^{k} Z_n^{(i)}$, where $Z_n^{(i)}$ is the number of individuals in the n-th generation of the subtree rooted at the i-th individual in the first generation. Hence

$$\mathbf{Q}_{(2.1)} = \frac{1}{m^{n+1}}\mathbf{P}(N = k) \sum_{i=1}^{k} \mathbf{E}\Big\{Z_n^{(i)} 1_{\{\mathbb{T}_1 \in A_1, \ldots, \mathbb{T}_k \in A_k\}} \Big| N = k\Big\}.$$

We have $\mathbf{P}(N = k) = p_k$, and

$$\mathbf{E}\{Z_n^{(i)}\,\mathbf{1}_{\{\mathbb{T}_1\in A_1,\dots,\,\mathbb{T}_k\in A_k\}} \mid N = k\} = \mathbf{E}[Z_n\,\mathbf{1}_{\{\mathbb{T}\in A_i\}}]\prod_{j\neq i}\mathbf{P}(A_j),$$

which is $m^n\,\mathbf{Q}(A_i)\prod_{j\neq i}\mathbf{P}(A_j)$. The lemma is proved.   □

It follows from Lemma 2.4 that the root $\varnothing$ of the size-biased Galton–Watson tree has the biased distribution, i.e., having k children with probability $\frac{kp_k}{m}$; among the individuals in the first generation, one of them is chosen randomly according to the uniform distribution: the subtree rooted at this vertex is a size-biased Galton–Watson tree, whereas the subtrees rooted at all other vertices in the first generation are usual Galton–Watson trees, and all these subtrees are independent.

We iterated the procedure, and obtain a decomposition of the size-biased Galton–Watson tree into an (infinite) spine and i.i.d. copies of the usual Galton–Watson tree: The root $\varnothing =: w_0$ has the biased distribution, i.e., having k children with probability $\frac{kp_k}{m}$. Among the children of the root, one of them is chosen randomly according to the uniform distribution, as the element of the spine in the first generation; let us denote this element by w_1. We attach subtrees rooted at all other children; they are independent copies of the usual Galton–Watson tree. The vertex w_1 has the biased distribution. Among the children of w_1, we choose at random one of them as the element of the spine in the second generation, denoted by w_2. Independent copies of the usual Galton–Watson tree are attached as subtrees rooted at all other children of w_1, whereas w_2 has the biased distribution. The system goes on indefinitely. See Fig. 2.3.

Having the application of the next section in mind, let us connect the size-biased Galton–Watson tree to the branching process with immigration. The latter starts with

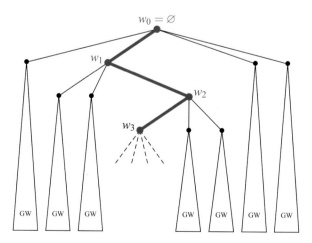

Fig. 2.3 A size-biased Galton–Watson tree

no individual (say), and is governed by a reproduction law and an immigration law. At generation n (for $n \geq 1$), Y_n new individuals are added into the system, while all individuals regenerate independently and following the same reproduction law; we assume that $(Y_n, \, n \geq 1)$ is a collection of i.i.d. random variables following the same immigration law, and independent of everything else up to that generation.

The size-biased Galton–Watson tree tells us that $(Z_n - 1, \, n \geq 0)$ under $\mathbf{Q}$ is a branching process with immigration, whose immigration law is that of $\widehat{N} - 1$, with $\mathbf{P}(\widehat{N} = k) := \frac{k p_k}{m}$, for $k \geq 1$.

2.3　Application: The Kesten–Stigum Theorem

We start with a dichotomy theorem for branching processes with immigration.

Theorem 2.5 (Seneta [219]) *Let Z_n be the number of individuals in the n-th generation of a branching process with immigration (Y_n). Assume that $1 < m < \infty$, where m denotes the expectation of the reproduction law.*

(i) *If $\mathbf{E}(\ln_+ Y_1) < \infty$, then $\lim_{n\to\infty} \frac{Z_n}{m^n}$ exists and is finite almost surely.*
(ii) *If $\mathbf{E}(\ln_+ Y_1) = \infty$, then $\limsup_{n\to\infty} \frac{Z_n}{m^n} = \infty$, a.s.*

Proof (ii) Assume $\mathbf{E}(\ln_+ Y_1) = \infty$. By the Borel–Cantelli lemma [102, Theorem 2.5.9], $\limsup_{n\to\infty} \frac{\ln Y_n}{n} = \infty$ a.s. Since $Z_n \geq Y_n$, it follows that for any $c > 1$, $\limsup_{n\to\infty} \frac{Z_n}{c^n} = \infty$, a.s.
(i) Assume now $\mathbf{E}(\ln_+ Y_1) < \infty$. By the law of large numbers, $\lim_{n\to\infty} \frac{\ln_+ Y_n}{n} = 0$ a.s., so for any $c > 0$, $\sum_k \frac{Y_k}{c^k} < \infty$ a.s.

Let $\mathcal{Y}$ be the σ-field generated by (Y_n). Clearly,

$$\mathbf{E}(Z_{n+1} \mid \mathscr{F}_n, \, \mathcal{Y}) = m Z_n + Y_{n+1} \geq m Z_n,$$

thus $(\frac{Z_n}{m^n})$ is a submartingale (conditionally on $\mathcal{Y}$), and $\mathbf{E}(\frac{Z_n}{m^n} \mid \mathcal{Y}) = \sum_{k=0}^{n} \frac{Y_k}{m^k}$. In particular, on the set $\{\sum_{k=0}^{\infty} \frac{Y_k}{m^k} < \infty\}$, we have $\sup_n \mathbf{E}(\frac{Z_n}{m^n} \mid \mathcal{Y}) < \infty$, so $\lim_{n\to\infty} \frac{Z_n}{m^n}$ exists and is finite. Since $\mathbf{P}(\sum_{k=0}^{\infty} \frac{Y_k}{m^k} < \infty) = 1$, the result follows. □

We recall an elementary result [102, Theorem 5.3.3]. Let $(\mathscr{F}_n)$ be a filtration, and let $\mathscr{F}_\infty$ be the smallest σ-field containing all $\mathscr{F}_n$. Let $\mathbf{P}$ and $\mathbf{Q}$ be probabilities on $(\Omega, \mathscr{F}_\infty)$. Assume that for any n, $\mathbf{Q}_{|\mathscr{F}_n} \ll \mathbf{P}_{|\mathscr{F}_n}$. Let $\xi_n := \frac{d\mathbf{Q}_{|\mathscr{F}_n}}{d\mathbf{P}_{|\mathscr{F}_n}}$, and let $\xi := \limsup_{n\to\infty} \xi_n$ which is $\mathbf{P}$-a.s. finite. Then

$$\mathbf{Q}(A) = \mathbf{E}(\xi \, \mathbf{1}_A) + \mathbf{Q}(A \cap \{\xi = \infty\}), \qquad \forall A \in \mathscr{F}_\infty.$$

It follows easily that

$$\mathbf{Q} \ll \mathbf{P} \Leftrightarrow \xi < \infty, \ \mathbf{Q}\text{-a.s.} \ \Leftrightarrow \ \mathbf{E}(\xi) = 1, \tag{2.2}$$

$$\mathbf{Q} \perp \mathbf{P} \Leftrightarrow \xi = \infty, \ \mathbf{Q}\text{-a.s.} \ \Leftrightarrow \ \mathbf{E}(\xi) = 0. \tag{2.3}$$

Proof of Theorem 2.3 If $\sum_{i=1}^{\infty} p_i\, i \ln i < \infty$, then $\mathbf{E}(\ln_+ \widehat{N}) < \infty$. By Theorem 2.5, $\lim_{n\to\infty} M_n$ exists $\mathbf{Q}$-a.s. and is finite $\mathbf{Q}$-a.s. In view of (2.2), this means $\mathbf{E}(M_\infty) = 1$; in particular, $\mathbf{P}(M_\infty = 0) < 1$, thus $\mathbf{P}(M_\infty = 0) = q$ (Lemma 2.2).

If $\sum_{i=1}^{\infty} p_i\, i \ln i = \infty$, then $\mathbf{E}(\ln_+ \widehat{N}) = \infty$. By Theorem 2.5, $\lim_{n\to\infty} M_n$ exists $\mathbf{Q}$-a.s. and is infinite $\mathbf{Q}$-a.s. Hence $\mathbf{E}(M_\infty) = 0$ (by (2.3)), i.e., $\mathbf{P}(M_\infty = 0) = 1$.

□

2.4 Notes

The material of this chapter is borrowed from Lyons et al. [176], and the presentation adapted from Chap. 1 of my lecture notes [221].

Section 2.1 collects a few elementary properties of Galton–Watson processes. For more detailed discussions, we refer to the books by Asmussen and Hering [31], Athreya and Ney [32], Harris [122].

The formalism described in Sect. 2.2 is due to Neveu [203]; the idea of viewing Galton–Watson trees as tree-valued random variables finds its root in Harris [122].

The technique of size-biased Galton–Watson trees, which goes back at least to Kahane and Peyrière [152], has been used by several authors in various contexts. Its presentation in Sect. 2.2, as well as its use to prove the Kesten–Stigum theorem, comes from Lyons et al. [176]. Size-biased Galton–Watson trees can actually be exploited to prove the corresponding results of the Kesten–Stigum theorem in the critical and subcritical cases. See [176] for more details.

Seneta's dichotomy theorem for branching processes with immigration (Theorem 2.5) was discovered by Seneta [219]; its short proof presented in Sect. 2.3 is borrowed from Asmussen and Hering [31, pp. 50–51].

Chapter 3
Branching Random Walks and Martingales

The Galton–Watson branching process counts the number of particles in each generation of a branching process. In this chapter, we produce an extension, in the spatial sense, by associating each individual of the branching process with a random variable. This results in a **branching random walk**. We present several martingales that are naturally related to the branching random walk, and study some elementary properties.

3.1 Branching Random Walks: Basic Notation

Let us briefly recall the definition of the branching random walk, introduced in Chap. 1: At time $n = 0$, one particle is at position 0. At time $n = 1$, the particle dies, giving birth to a certain number of children distributed according to a given point process Ξ. At time $n = 2$, all these particles die, each producing children positioned (with respect to their birth places) according to the same point process Ξ, independently of each other and of everything up to then. The system goes on indefinitely as long as there are particles alive.

Let $\mathbb{T}$ denote the genealogical tree of the system, and $(V(x), x \in \mathbb{T})$ the positions of the individuals in the system. As before, $|x|$ stands for the generation of x, and x_i (for $0 \le i \le |x|$) for the ancestor of x in the i-th generation. We write $[\![\varnothing, x]\!] := \{x_0 := \varnothing, x_1, \ldots, x_{|x|}\}$ to denote the set of vertices (including $\varnothing$ and x) in the unique shortest path connecting the root $\varnothing$ to x.

For two vertices x and y of $\mathbb{T}$, we write $x < y$ (or $y > x$) if y is a descendant of x, and $x \le y$ (or $y \ge x$) if either $x < y$ or $x = y$.

For any $x \in \mathbb{T}\backslash\{\varnothing\}$, we denote by $\overleftarrow{x}$ its parent, and by $\mathrm{brot}(x)$ the set of the brothers of x, which can be possibly empty; so $y \in \mathrm{brot}(x)$ indicates y is different from x but having the same parent as x.

© Springer International Publishing Switzerland 2015
Z. Shi, *Branching Random Walks*, Lecture Notes in Mathematics 2151,
DOI 10.1007/978-3-319-25372-5_3

As before, the (log-)Laplace transform of the point process Ξ plays an important role:

$$\psi(\beta) := \ln \mathbf{E}\Big(\sum_{x\in\mathbb{T}: |x|=1} e^{-\beta V(x)} \Big) = \ln \mathbf{E}\Big(\sum_{u\in\Xi} e^{-\beta u} \Big) \in (-\infty, \infty], \quad \beta \in \mathbb{R},$$

where $|x| = 1$ indicates that x is in the first generation of the branching random walk. We regularly write $\sum_{|x|=1}(\cdots)$ instead of $\sum_{x\in\mathbb{T}: |x|=1}(\cdots)$.

We always assume that $\psi(0) > 0$. The genealogical tree $\mathbb{T}$ is a Galton–Watson process (often referred to as the associated or underlying Galton–Watson process), which is supercritical under the assumption $\psi(0) > 0$. In particular, according to Theorem 2.1 in Sect. 2.1, our system survives with positive probability.

Quite frequently, we are led to work on the set of non-extinction, so it is convenient to introduce the new probability

$$\mathbf{P}^*(\,\cdot\,) := \mathbf{P}(\,\cdot\,|\, \text{non-extinction}).$$

We close this section with the following result.

Lemma 3.1 *Assume that $\psi(0) > 0$. If $\psi(t) \le 0$ for some $t > 0$, then[1]*

$$\lim_{n\to\infty} \inf_{|x|=n} V(x) = \infty, \quad \mathbf{P}^*\text{-}a.s.$$

Proof Let $t > 0$ be such that $\psi(t) \le 0$. Without loss of generality, we assume $t = 1$ (otherwise, we consider $tV(x)$ in place of $V(x)$). Let

$$W_n := \sum_{|x|=n} e^{-n\psi(1)-V(x)}, \quad n \ge 0,$$

which is a non-negative martingale, so it converges, when $n \to \infty$, to a non-negative random variable, say W_∞. We have $\mathbf{E}[W_\infty] \le 1$ by Fatou's lemma.

Let $Y := \limsup_{n\to\infty} e^{-\inf_{|x|=n} V(x)}$. Since $e^{-\inf_{|x|=n} V(x)} \le \sum_{|x|=n} e^{-V(x)} \le W_n$ (recalling that $\psi(1) \le 0$ by assumption), we have $\mathbf{E}(Y) \le \mathbf{E}[W_\infty] \le 1$.

It remains to check that $Y = 0$ a.s. (which is equivalent to saying that $Y = 0$ $\mathbf{P}^*$-a.s.), or equivalently, $\liminf_{n\to\infty} \inf_{|x|=n} V(x) = \infty$ a.s.

Looking at the subtrees rooted at each of the vertices in the first generation, we immediately get

$$Y = \sup_{|x|=1} [e^{-V(x)} Y(x)],$$

---

[1]If $\mathbf{P}\{\sum_{|x|=1} \mathbf{1}_{\{V(x)>0\}} > 0\} > 0$, then the condition that $\psi(t) \le 0$ for some $t > 0$ is also necessary to have $\inf_{|x|=n} V(x) \to \infty$, $\mathbf{P}^*$-a.s. See Biggins [52].

where $(Y(x))$ are independent copies of Y, and independent of $(V(x), |x| = 1)$ given $(x, |x| = 1)$. In particular, $\mathbf{E}(Y) = \mathbf{E}[\sup_{|x|=1} e^{-V(x)} Y(x)]$.

The system is supercritical by assumption, so with positive probability, the maximum expression $\sup_{|x|=1} e^{-V(x)} Y(x)$ involves at least two terms. Thus, if $\mathbf{E}(Y) > 0$, then we would have

$$\mathbf{E}\Big[\sup_{|x|=1} e^{-V(x)} Y(x) \Big] < \mathbf{E}\Big[\sum_{|x|=1} e^{-V(x)} Y(x) \Big] = \mathbf{E}\Big[\sum_{|x|=1} e^{-V(x)} \Big] \mathbf{E}(Y) = e^{\psi(1)} \mathbf{E}(Y),$$

which would lead to a contradiction because $\psi(1) \le 0$ by assumption. Therefore, $Y = 0$ a.s.

$\square$

3.2 The Additive Martingale

Assume $\psi(1) < \infty$. Let

$$W_n := \sum_{|x|=n} e^{-n\psi(1)-V(x)}, \quad n \ge 0.$$

Clearly, $(W_n, \; n \ge 0)$ is a martingale with respect to the natural filtration of the branching random walk, and is called an *additive martingale* [204].

Since W_n is a non-negative martingale, we have

$$W_n \to W_\infty, \quad \text{a.s.},$$

for some non-negative random variable W_∞. Fatou's lemma says that $\mathbf{E}(W_\infty) \le 1$. An important question is whether the limit W_∞ is degenerate. By an argument as in the proof of Lemma 2.2 of Sect. 2.1, we can check [55] that $\mathbf{P}\{W_\infty = 0\}$ is either q or 1. Therefore, $\mathbf{P}\{W_\infty > 0\} > 0$ is equivalent to saying that $W_\infty > 0$, $\mathbf{P}^*$-a.s., and also means $\mathbf{P}\{W_\infty = 0\} = \mathbf{P}\{\text{extinction}\}$.

Here is Biggins's martingale convergence theorem, which is a spatial extension of the Kesten–Stigum theorem. We write

$$\psi'(1) := -\mathbf{E}\Big[\sum_{|x|=1} V(x) e^{-\psi(1)-V(x)} \Big],$$

whenever $\mathbf{E}[\sum_{|x|=1} |V(x)| e^{-V(x)}] < \infty$, and we simply say "if $\psi'(1) \in \mathbb{R}$". A similar remark applies to the forthcoming "$\psi'(\beta) \in \mathbb{R}$".

Theorem 3.2 (Biggins Martingale Convergence Theorem) *Assume $\psi(0) > 0$. If $\psi(1) < \infty$ and $\psi'(1) \in \mathbb{R}$, then*

$$\mathbf{E}(W_\infty) = 1 \iff W_\infty > 0, \; \mathbf{P}^*\text{-a.s.}$$

$$\iff \mathbf{E}(W_1 \ln_+ W_1) < \infty \;\text{ and }\; \psi(1) > \psi'(1).$$

Proof Postponed to Sect. 4.8. $\square$

For any $\beta \in \mathbb{R}$ with $\psi(\beta) < \infty$, by considering βV instead of V, the Biggins theorem has the following general form: Let $\beta \in \mathbb{R}$ be such that $\psi(\beta) < \infty$, and let $W_n^{(\beta)} := \sum_{|x|=n} e^{-n\psi(\beta)-\beta V(x)}$ which is a non-negative martingale and which converges a.s. to, say, $W_\infty^{(\beta)}$.

Theorem 3.3 (Biggins [50]) *Assume $\psi(0) > 0$. Let $\beta \in \mathbb{R}$ be such that $\psi(\beta) < \infty$ and that $\psi'(\beta) := -\mathbf{E}\{\sum_{|x|=1} V(x)e^{-\psi(\beta)-\beta V(x)}\} \in \mathbb{R}$, then*

$$\mathbf{E}[W_\infty^{(\beta)}] = 1 \iff \mathbf{P}(W_\infty^{(\beta)} = 0) < 1$$

$$\iff \mathbf{E}[W_1^{(\beta)} \ln_+ W_1^{(\beta)}] < \infty \ \text{and} \ \beta\psi'(\beta) < \psi(\beta).$$

Theorem 3.3 reduces to the Kesten–Stigum theorem (Theorem 2.3 in Sect. 2.1) when $\beta = 0$, and is equivalent to Theorem 3.2 if $\beta \neq 0$.

3.3   The Multiplicative Martingale

Let $(V(x))$ be a branching random walk such that $\psi(0) > 0$. The basic assumption in this section is: $\psi(1) = 0$, $\psi'(1) \leq 0$.[2]

Assume that $\Phi(s) := \mathbf{E}(e^{-s\xi^*})$, $s \geq 0$, for some non-negative random variable ξ^* with $\mathbf{P}(\xi^* > 0) > 0$ (so $\Phi(s) < 1$ for any $s > 0$), is such that[3]

$$\Phi(s) = \mathbf{E}\left[\prod_{|x|=1} \Phi(se^{-V(x)}) \right], \quad \forall s \geq 0. \tag{3.1}$$

[Notation: $\prod_\varnothing := 1$.] For the existence[4] of such a function Φ under our assumption ($\psi(0) > 0$, $\psi(1) = 0$ and $\psi'(1) \leq 0$), see Liu [167].

[An equivalent way to state (3.1) is as follows: ξ^* has the same distribution as $\sum_{|x|=1} \xi_x^* e^{-V(x)}$, where (ξ_x^*) are independent copies of ξ^*, independent of $(V(x))$.]

For any $t > 0$, let

$$M_n^{(t)} := \prod_{|x|=n} \Phi(te^{-V(x)}), \quad n \geq 0,$$

---

[2]It is always possible, by means of a simple translation, to make a branching random walk satisfy $\psi(1) = 0$ as long as $\psi(1) < \infty$. The condition $\psi'(1) \leq 0$ is more technical: It is to guarantee the existence of the forthcoming function Φ; see the paragraph below.

[3]In Proposition 3.4 and Lemma 3.5 below, we simply say that Φ is a Laplace transform satisfying (3.1).

[4]In fact, it is also unique, up to a multiplicative constant in the argument. See Biggins and Kyprianou [56].

which is a martingale, called *multiplicative martingale* [204]. Since $M_n^{(t)}$, taking values in [0, 1], is bounded, there exists a random variable $M_\infty^{(t)} \in [0, 1]$ such that

$$M_n^{(t)} \to M_\infty^{(t)}, \quad \text{a.s.},$$

and in L^p for any $1 \le p < \infty$. In particular, $\mathbf{E}[M_\infty^{(t)}] = \Phi(t)$.

Let us collect a few elementary properties of the limiting random variable $M_\infty^{(t)}$.

Proposition 3.4 *Assume* $\psi(0) > 0$, $\psi(1) = 0$ *and* $\psi'(1) \le 0$. *Let* Φ *be a Laplace transform satisfying* (3.1). *Then*

(i) $M_\infty^{(t)} = [M_\infty^{(1)}]^t$, $\forall t > 0$.
(ii) $M_\infty^{(1)} > 0$ a.s.
(iii) $M_\infty^{(1)} < 1$, $\mathbf{P}^*$-a.s.
(iv) $\ln \frac{1}{M_\infty^{(1)}}$ *has Laplace transform* Φ.

The proof of Proposition 3.4 relies on the following result. A function L is said to be slowly varying at 0 if for any $a > 0$, $\lim_{s \to 0} \frac{L(as)}{L(s)} = 1$.

Lemma 3.5 *Assume* $\psi(0) > 0$, $\psi(1) = 0$ *and* $\psi'(1) \le 0$. *Let* Φ *be a Laplace transform satisfying* (3.1). *Then the function*

$$L(s) := \frac{1 - \Phi(s)}{s} > 0, \quad s > 0,$$

is slowly varying at 0.

Proof of Lemma 3.5 Assume L is not slowly varying at 0. So there would be $0 < a < 1$ and a sequence (s_k) with $s_k \downarrow 0$ such that $\frac{L(s_k a)}{L(s_k)} \to b \neq 1$. By integration by parts, L is also the Laplace transform of a measure on $[0, \infty)$; so the function $s \mapsto L(s)$ is non-increasing. In particular, $b > 1$, and for any $a' \in (0, a]$,

$$\liminf_{k \to \infty} \frac{L(a' s_k)}{L(s_k)} \ge b > 1.$$

On the other hand, writing $x^{(1)}, x^{(2)}, \ldots, x^{(Z_n)}$ for the vertices in the n-th generation, then for any $s > 0$,

$$L(s) = \mathbf{E}\left[s^{-1} \left(1 - \prod_{|x|=n} \Phi(se^{-V(x)}) \right) \right] \quad \text{(by (3.1))}$$

$$= \mathbf{E}\left[\sum_{j=1}^{Z_n} \frac{1 - \Phi(se^{-V(x^{(j)})})}{s} \prod_{i=1}^{j-1} \Phi(se^{-V(x^{(i)})}) \right] \quad (\prod_\emptyset := 1)$$

$$= \mathbf{E}\left[\sum_{j=1}^{Z_n} e^{-V(x^{(j)})} L(se^{-V(x^{(j)})}) \prod_{i=1}^{j-1} \Phi(se^{-V(x^{(i)})}) \right], \quad (\tfrac{1-\Phi(r)}{r} = L(r)) \quad (3.2)$$

i.e.,

$$1 = \mathbf{E}\Big[\sum_{j=1}^{Z_n} e^{-V(x^{(j)})}\, \frac{L(se^{-V(x^{(j)})})}{L(s)} \prod_{i=1}^{j-1} \Phi(se^{-V(x^{(i)})})\Big].$$

For $s = s_k$, by Fatou's lemma,

$$1 \geq \mathbf{E}\Big[\sum_{j=1}^{Z_n} e^{-V(x^{(j)})}\, \liminf_{k\to\infty} \frac{L(s_k e^{-V(x^{(j)})})}{L(s_k)} \prod_{i=1}^{j-1} \Phi(s_k e^{-V(x^{(i)})})\Big].$$

Since Φ is continuous with $\Phi(0) = 1$, this yields

$$1 \geq \mathbf{E}\Big[\sum_{j=1}^{Z_n} e^{-V(x^{(j)})} \Big(b\, \mathbf{1}_{\{e^{-V(x^{(j)})}\leq a\}} + \mathbf{1}_{\{a < e^{-V(x^{(j)})}\leq 1\}}\Big)\Big]$$

$$= \mathbf{E}\Big[\sum_{|x|=n} e^{-V(x)} \Big(b\, \mathbf{1}_{\{e^{-V(x)}\leq a\}} + \mathbf{1}_{\{a < e^{-V(x)}\leq 1\}}\Big)\Big]$$

$$= (b-1)\mathbf{E}\Big[\sum_{|x|=n} e^{-V(x)} \mathbf{1}_{\{e^{-V(x)}\leq a\}}\Big] + \mathbf{E}\Big[\sum_{|x|=n} e^{-V(x)} \mathbf{1}_{\{e^{-V(x)}\leq 1\}}\Big].$$

Since $\mathbf{E}[\sum_{|x|=n} e^{-V(x)}] = e^{\psi(1)} = 1$, this means:

$$\mathbf{E}\Big[\sum_{|x|=n} e^{-V(x)} \mathbf{1}_{\{e^{-V(x)}>1\}}\Big] \geq (b-1)\mathbf{E}\Big[\sum_{|x|=n} e^{-V(x)} \mathbf{1}_{\{e^{-V(x)}\leq a\}}\Big].$$

Applying the many-to-one formula (Theorem 1.1 in Sect. 1.3) to $t = 1$ gives $\mathbf{P}\{e^{-S_n} > 1\} \geq (b-1)\mathbf{P}\{e^{-S_n} \leq a\}$, i.e.,

$$\mathbf{P}\{S_n < 0\} \geq (b-1)\mathbf{P}\{S_n \geq -\ln a\}.$$

If $\psi'(1) < 0$, then $\mathbf{E}(S_1) = -\psi'(1) > 0$ whereas $0 < a < 1$, we have $\frac{\mathbf{P}\{S_n<0\}}{\mathbf{P}\{S_n\geq -\ln a\}} \to 0, n \to \infty$. Thus $b \leq 1$, which contradicts the assumption $b > 1$. As a consequence, L is slowly varying at 0 in case $\psi'(1) < 0$.

It remains to treat the case $\psi'(1) = 0$. Consider the sequence of functions $(f_k, k \geq 1)$ on $[0, \infty)$ defined by $f_k(y) := \exp(-\frac{L(s_k y)}{L(s_k)})$, $y \geq 0$. Since each f_k takes values in $[0, 1]$ and is non-decreasing, Helly's selection principle [157, p. 372, Theorem 5][5] says that there exists a subsequence of (s_k), still denoted by (s_k) by

---

[5]In [157], Helly's selection principle is stated for functions on a compact interval. We apply it to each of the intervals $[0, n]$, and then conclude by a diagonal argument (i.e., taking the diagonal elements in a double array).

an abuse of notation, such that for all $y > 0$, $\exp(-\frac{L(s_k y)}{L(s_k)})$ converges to a limit, say $e^{-g(y)}$.

By (3.2) (and Fatou's lemma as before), for any $y > 0$,

$$g(y) \geq \mathbf{E}\left[\sum_{j=1}^{Z_n} e^{-V(x^{(j)})} g(ye^{-V(x^{(j)})})\right] = \mathbf{E}\left[\sum_{|x|=1} e^{-V(x)} g(ye^{-V(x)})\right],$$

which is $\mathbf{E}[g(ye^{-S_1})]$ by the many-to-one formula (Theorem 1.1 in Sect. 1.3). This implies that $(g(ye^{-S_n}), n \geq 0)$ is a non-negative supermartingale, which converges a.s. to, say G_y. Since $\mathbf{E}(S_1) = -\psi'(1) = 0$, we have $\limsup_{n\to\infty} e^{-S_n} = \infty$ a.s., and $\liminf_{n\to\infty} e^{-S_n} = 0$ a.s. So by monotonicity of g, $g(\infty) = G_y = g(0+)$: g is a constant. Since $g(a) = b > 1 = g(1)$, this leads to a contradiction. $\qquad\square$

Proof of Proposition 3.4 (i) We claim that

$$\sum_{|x|=n} [\Phi(te^{-V(x)}) - 1] \to \ln M_\infty^{(t)}, \quad \mathbf{P}^*\text{-a.s.} \tag{3.3}$$

Indeed, since $u - 1 \geq \ln u$ for any $u \in (0, 1]$, we have, $\mathbf{P}^*$-almost surely,

$$\sum_{|x|=n} [\Phi(te^{-V(x)}) - 1] \geq \sum_{|x|=n} \ln \Phi(te^{-V(x)}) = \ln M_n^{(t)} \to \ln M_\infty^{(t)},$$

giving the lower bound in (3.3). For the upper bound, let $\varepsilon > 0$. By Lemma 3.1 (Sect. 3.1), $\inf_{|x|=n} V(x) \to \infty$ $\mathbf{P}^*$-a.s., so for $\mathbf{P}^*$-almost surely all sufficiently large n, $\Phi(te^{-V(x)}) - 1 \leq (1 - \varepsilon) \ln \Phi(te^{-V(x)})$, for all x with $|x| = n$, which leads to:

$$\sum_{|x|=n} [\Phi(te^{-V(x)}) - 1] \leq (1 - \varepsilon) \sum_{|x|=n} \ln \Phi(te^{-V(x)}),$$

and the latter converges to $(1 - \varepsilon) \ln M_\infty^{(t)}$, $\mathbf{P}^*$-a.s. This justifies (3.3).

Recall that $\Phi(s) - 1 = sL(s)$. Thus, on the set of non-extinction,

$$\frac{1}{t}\frac{\sum_{|x|=n}[\Phi(te^{-V(x)}) - 1]}{\sum_{|x|=n}[\Phi(e^{-V(x)}) - 1]} - 1 = \sum_{|x|=n}\frac{\Phi(e^{-V(x)}) - 1}{\sum_{|y|=n}\Phi(e^{-V(y)}) - 1}\left(\frac{L(te^{-V(x)})}{L(e^{-V(x)})} - 1\right).$$

We now look at the expressions on the left- and right-hand sides: Since L is slowly varying at 0 (Lemma 3.5), whereas $\inf_{|x|=n} V(x) \to \infty$ $\mathbf{P}^*$-a.s. (Lemma 3.1 of Sect. 3.1), thus the expression on the right-hand side tends to 0 $\mathbf{P}^*$-almost surely; the expression on the left-hand side converges $\mathbf{P}^*$-almost surely to $\frac{1}{t}\frac{\ln M_\infty^{(t)}}{\ln M_\infty^{(1)}}$ (see (3.3)).

Therefore, $\mathbf{P}^*$-a.s.,

$$\frac{1}{t}\frac{\ln M_\infty^{(t)}}{\ln M_\infty^{(1)}} = 0,$$

i.e., $M_\infty^{(t)} = [M_\infty^{(1)}]^t$, $\mathbf{P}^*$-a.s. Since $M_\infty^{(t)} = 1$ (for all t) on the set of extinction, we have $M_\infty^{(t)} = [M_\infty^{(1)}]^t$.

(iv) By (i), we have $\mathbf{E}[M_\infty^{(t)}] = \mathbf{E}\{[M_\infty^{(1)}]^t\}$. On the other hand, we have already seen that $\mathbf{E}[M_\infty^{(t)}] = \Phi(t)$. Thus $\Phi(t) = \mathbf{E}\{[M_\infty^{(1)}]^t\}$, $\forall t > 0$; thus Φ is the Laplace transform of $-\ln M_\infty^{(1)}$.

(ii) By assumption, Φ is the Laplace transform of a non-degenerate random variable, so (ii) follows from (iv).

(iii) By definition (3.1) of Φ, we have $\Phi(\infty) = \lim_{s\to\infty} \mathbf{E}[\prod_{|x|=1} \Phi(se^{-V(x)})]$, which, by dominated convergence, is $\mathbf{E}[\Phi(\infty)^N]$, where $N := \sum_{|x|=1} 1$. Therefore, $\Phi(\infty)$ satisfies $\Phi(\infty) = f(\Phi(\infty))$, with f denoting the generating function of N. By Theorem 2.1 in Sect. 2.1, $\Phi(\infty)$ is either $\mathbf{P}\{\text{extinction}\}$, or 1.

By definition, Φ is the Laplace transform of ξ^* with $\mathbf{P}\{\xi^* > 0\} > 0$; so $\Phi(\infty) < 1$, which means that $\Phi(\infty) = \mathbf{P}\{\text{extinction}\}$. On the other hand, (iv) tells us that $\mathbf{P}\{M_\infty^{(1)} = 1\} = \Phi(\infty)$. So $\mathbf{P}\{M_\infty^{(1)} = 1\} = \mathbf{P}\{\text{extinction}\}$. Since $\{M_\infty^{(1)} = 1\}$ contains the set of extinction, the two sets coincide almost surely.   □

3.4 The Derivative Martingale

Assuming $\psi(1) = 0$ and $\psi'(1) = 0$, we see that

$$D_n := \sum_{|x|=n} V(x)e^{-V(x)}, \quad n \geq 0,$$

is a martingale, called the *derivative martingale*.

The derivative martingale is probably *the* most important martingale associated with branching random walks. We postpone our study of (D_n) to Chap. 5. In particular, we are going to see, in Theorem 5.2 (Sect. 5.2), that under some general assumptions upon the law of the branching random walk, D_n converges a.s. to a *non-negative* limit. Furthermore, this non-negative limit is shown, in Theorem 5.29 (Sect. 5.6), to be closely related to the limit of the additive martingale, after a suitable normalisation.

Section 5.4 will reveal a crucial role played by the derivative martingale in the study of extreme values in branching random walks.

3.5 Notes

Most of the material in this chapter can be found in Biggins and Kyprianou [58].

Lemma 3.1 in Sect. 3.1 is due to Liu [167] and Biggins [52]. The main idea of our proof is borrowed from Biggins [52].

The Biggins martingale convergence theorem (Theorem 3.2 in Sect. 3.2) is originally proved by Biggins [50] under a slightly stronger condition. The theorem under the current condition can be found in Lyons [173].

In the case where the limit W_∞ in the Biggins martingale convergence theorem is non-degenerate, it is interesting to study its law. See, for example, Biggins and Grey [55] and Liu [169] for absolute continuity, and Liu [168] and Buraczewski [80] for precise tail estimates.

If the almost sure convergence of the non-negative martingale $(W_n^{(\beta)})$ is obvious for any given $\beta \in \mathbb{R}$ (such that $\psi(\beta) < \infty$), it is far less obvious whether or not the convergence holds uniformly in β. This problem is dealt with by Biggins [51]. The rate of convergence for the additive martingale is studied by several authors; see for example Iksanov and Meiners [145], Iksanov and Kabluchko [144].

The importance of multiplicative martingales studied in Sect. 3.3 is stressed by Neveu [204]. These martingales are defined in terms of solution of Eq. (3.1). The study of existence and uniqueness of (3.1) has a long history, going back at least to Kesten and Stigum [155], and has since been a constant research topic in various contexts (fixed points of smoothing transforms, stochastically self-similar fractals, multiplicative cascades, etc). Early contributions are from Doney [98], Mandelbrot [193], Kahane and Peyrière [152], Biggins [50], Holley and Liggett [130], Durrett and Liggett [103], Mauldin and Williams [194], Falconer [106], Guivarc'h [119], to name but a few. The 1990–2000 decade saw results established in the generality we are interested in, almost simultaneously by Liu [167], Biggins and Kyprianou [56], Lyons [173], Kyprianou [160]. We refer to [58] for a detailed survey, as well as to Alsmeyer et al. [22] and Buraczewski et al. [81] together with the references therein for recent extensions in various directions.

Multiplicative martingales are also particularly useful in the study of branching Brownian motion and the F-KPP equation; see the lecture notes of Berestycki [43].

The proof of Lemma 3.5 in case $\psi'(1) < 0$ is borrowed from Biggins and Kyprianou [56], and in case $\psi'(1) = 0$ from Kyprianou [160].

Derivative martingales, introduced in Sect. 3.4, are studied for branching Brownian motion by Lalley and Sellke [164], Neveu [204], Kyprianou [162], and for the branching random walk by Liu [168], Kyprianou [160], Biggins and Kyprianou [57], Aïdékon [8]. Although not mentioned here, it is closely related to multiplicative martingales; see Liu [168], and Harris [123] in the setting of branching Brownian motion.

Joffe [149] studies another interesting martingale naturally associated with the branching random walk in the case of i.i.d. random variables attached to edges of a Galton–Watson tree.

The martingales considered in this chapter are sums, or products, over particles in a same generation. Just as important as considering stopping times in martingale theory, it is often interesting to consider sums over particles belonging to some special random collections, called stopping lines. The basic framework is set up in Jagers [148] and Chauvin [84]; for a sample of interesting applications, see Biggins and Kyprianou [56–58], Kyprianou [161], Maillard [186], Olofsson [205].

Chapter 4
The Spinal Decomposition Theorem

This chapter is devoted to an important tool in the study of branching random walks: the **spinal decomposition**. In particular, it gives a probabilistic explanation for the presence of the one-dimensional random walk (S_n) appearing in the many-to-one formula (Theorem 1.1 in Sect. 1.3). We establish a general spinal decomposition theorem for branching random walks. In order to do so, we need to introduce the notions of spines and changes of probabilities, which are the main topics of the first two sections. Two special cases of the spinal decomposition theorem are particularly useful; they are presented, respectively, in Example 4.5 (Sect. 4.6) for the size-biased branching random walk, and in Example 4.6 (Sect. 4.7) where the branching random walk is above a given level along the spine. The power of the spinal decomposition theorem will be seen via a few case studies in the following chapters. Here, we prove in Sect. 4.8, as a first application, the Biggins martingale convergence theorem for the branching random walk, already stated in Sect. 3.2 as Theorem 3.2.

4.1 Attaching a Spine to the Branching Random Walk

The spinal decomposition theorem describes the distribution of the paths of the branching random walk. This description is formulated by means of a particular infinite ray (see below for details)—called spine—on the associated Galton–Watson tree, and of a new probability. For the sake of clarity, we present spinal decompositions via three steps. In the first step, the notion of spine is introduced. In the second step, we construct a new probability. In the third and last step, the spinal decomposition theorem is presented.

The branching random walk $V := (V(x), x \in \mathbb{T})$ can be considered as a random variable taking values in the space of marked trees, while the associated supercritical Galton–Watson tree $\mathbb{T}$ is a random variable taking values in the space of rooted trees. We now attach to $(V(x), x \in \mathbb{T})$ an additional random infinite ray $w = (w_n, n \geq 0)$,

© Springer International Publishing Switzerland 2015
Z. Shi, *Branching Random Walks*, Lecture Notes in Mathematics 2151,
DOI 10.1007/978-3-319-25372-5_4

called **spine**. By an infinite ray (sometimes also referred to as an infinite path), we mean $w_0 := \varnothing$ and $\overleftarrow{w_n} = w_{n-1}$ (recalling that $\overleftarrow{x}$ is the parent of x) for any $n \geq 1$, i.e., each w_n is a child of w_{n-1}. In particular, $|w_n| = n$, $\forall n \geq 0$.

In the rest of the chapter, for $n \geq 0$, we write

$$\mathscr{F}_n := \sigma\{V(x), \ x \in \mathbb{T}, \ |x| \leq n\},$$

which is the σ-field generated by the branching random walk in the first n generations. Let

$$\mathscr{F}_\infty := \sigma\{V(x), \ x \in \mathbb{T}\},$$

which contains all the information given by the branching random walk. In general, the spine w is not $\mathscr{F}_\infty$-measurable: There is extra randomness in w.

4.2 Harmonic Functions and Doob's *h*-Transform

Let $(V(x))$ be a branching random walk such that $\mathbf{E}[\sum_{|x|=1} e^{-V(x)}] = 1$. Let $(S_n - S_{n-1}, \ n \geq 1)$ be a sequence of i.i.d. real-valued random variables; the law of $S_1 - S_0$ is as follows: for any Borel function $g : \mathbb{R} \to [0, \infty)$,

$$\mathbf{E}[g(S_1 - S_0)] = \mathbf{E}\left[\sum_{|x|=1} g(V(x)) e^{-V(x)} \right]. \tag{4.1}$$

For $a \in \mathbb{R}$, let $\mathbf{P}_a$ denote the probability such that $\mathbf{P}_a(S_0 = a) = 1$ and that $\mathbf{P}_a(V(\varnothing) = a) = 1$, and $\mathbf{E}_a$ the expectation with respect to $\mathbf{P}_a$.[1] We often refer to (S_n) as an **associated random walk**.

Let $\mathrm{D} \subset \mathbb{R}$ be a Borel set of $\mathbb{R}$, such that[2]

$$\mathbf{P}_a(S_1 \in \mathrm{D}) > 0, \quad \forall a \in \mathrm{D}. \tag{4.2}$$

Let $h : \mathrm{D} \to (0, \infty)$ be a positive **harmonic function** associated with (S_n), i.e.,

$$h(a) = \mathbf{E}_a[h(S_1) \mathbf{1}_{\{S_1 \in \mathrm{D}\}}], \quad \forall a \in \mathrm{D}. \tag{4.3}$$

---

[1] If $a = 0$, we write $\mathbf{P}$ and $\mathbf{E}$ in place of $\mathbf{P}_0$ and $\mathbf{E}_0$, respectively. A similar remark applies to the forthcoming probabilities $\mathbf{Q}_a^{(h)}$ and $\mathbf{Q}_a$.

[2] Very often, we take $\mathrm{D} := \mathbb{R}$, in which case (4.2) is automatically satisfied.

We now define the random walk (S_n) (under $\mathbf{P}_a$) conditioned to stay in D, in the sense of Doob's h-transform: it is a Markov chain with transition probabilities given by

$$p^{(h)}(u, \, dv) := \mathbf{1}_{\{v \in \mathrm{D}\}} \frac{h(v)}{h(u)} \mathbf{P}_u(S_1 \in dv), \quad u \in \mathrm{D}. \tag{4.4}$$

4.3   Change of Probabilities

Assume $\psi(1) = 0$, i.e., $\mathbf{E}[\sum_{|x|=1} e^{-V(x)}] = 1$. Let (S_n) be an associated random walk in the sense of (4.1).

Let $\mathrm{D} \subset \mathbb{R}$ be a Borel set satisfying (4.2), and let $a \in \mathrm{D}$. Let $h : \mathrm{D} \to (0, \infty)$ be a positive harmonic function in the sense of (4.3). Define

$$M_n^{(h)} := \sum_{|x|=n} h(V(x)) e^{-V(x)} \mathbf{1}_{\{V(y) \in \mathrm{D}, \, \forall y \in [\![\varnothing, \, x]\!]\}}, \quad n \geq 0, \tag{4.5}$$

where $[\![\varnothing, \, x]\!]$ denotes, as before, the set of vertices in the unique shortest path connecting the root $\varnothing$ to x. We mention that $M_n^{(h)}$ has nothing to do with the multiplicative martingale studied in Sect. 3.3.

Lemma 4.1 *Let $a \in \mathrm{D}$. The process $(M_n^{(h)}, \, n \geq 0)$ is a martingale with respect to the expectation $\mathbf{E}_a$ and to the filtration $(\mathscr{F}_n)$.*

Proof By definition,

$$M_{n+1}^{(h)} = \sum_{|z|=n} \; \sum_{x: \, |x|=n+1, \, \overleftarrow{x}=z} h(V(x)) e^{-V(x)} \mathbf{1}_{\{V(y) \in \mathrm{D}, \, \forall y \in [\![\varnothing, x]\!]\}}$$

$$= \sum_{|z|=n} \mathbf{1}_{\{V(y) \in \mathrm{D}, \, \forall y \in [\![\varnothing, z]\!]\}} \sum_{x: \, |x|=n+1, \, \overleftarrow{x}=z} h(V(x)) e^{-V(x)} \mathbf{1}_{\{V(x) \in \mathrm{D}\}}.$$

Therefore,

$$\mathbf{E}_a(M_{n+1}^{(h)} \mid \mathscr{F}_n) = \sum_{|z|=n} \mathbf{1}_{\{V(y) \in \mathrm{D}, \, \forall y \in [\![\varnothing, z]\!]\}} \mathbf{E}_{V(z)} \Big(\sum_{|x|=1} e^{-V(x)} h(V(x)) \mathbf{1}_{\{V(x) \in \mathrm{D}\}} \Big).$$

By the many-to-one formula (Theorem 1.1 in Sect. 1.3), for any $b \in \mathrm{D}$,

$$\mathbf{E}_b \Big(\sum_{|x|=1} h(V(x)) e^{-V(x)} \mathbf{1}_{\{V(x) \in \mathrm{D}\}} \Big) = e^{-b} \mathbf{E}_b \Big(h(S_1) \mathbf{1}_{\{S_1 \in \mathrm{D}\}} \Big)$$

$$= e^{-b} h(b). \quad \text{(by (4.3))}$$

As a consequence, $\mathbf{E}_a(M_{n+1}^{(h)} \mid \mathscr{F}_n) = M_n^{(h)}$. $\qquad\qquad\qquad\qquad\qquad\qquad\qquad \square$

Since $(M_n^{(h)}, n \geq 0)$ is a non-negative martingale with $\mathbf{E}_a(M_n^{(h)}) = h(a)e^{-a}$, for all n, it follows from Kolmogorov's extension theorem that there exists a unique probability measure $\mathbf{Q}_a^{(h)}$ on $\mathscr{F}_\infty$ such that

$$\mathbf{Q}_a^{(h)}(A) = \int_A \frac{M_n^{(h)}}{h(a)e^{-a}}\, d\mathbf{P}_a, \quad \forall A \in \mathscr{F}_n, \ \forall n \geq 0. \tag{4.6}$$

In words, $\frac{M_n^{(h)}}{h(a)e^{-a}}$ is the Radon–Nikodym derivative with respect to the restriction of $\mathbf{P}_a$ on $\mathscr{F}_n$, of the restriction of $\mathbf{Q}_a^{(h)}$ on $\mathscr{F}_n$.

We end this section with the following simple result which is not needed in establishing the spinal decomposition theorem in the next sections, but which is sometimes useful in the applications of the theorem.

Note that $M_n^{(h)} > 0$, $\mathbf{Q}_a^{(h)}$-a.s. Note that the $\mathbf{P}_a$-martingale $(M_n^{(h)}, n \geq 0)$ being non-negative, there exists $M_\infty^{(h)} \geq 0$ such that $M_n^{(h)} \to M_\infty^{(h)}$, $\mathbf{P}_a$-a.s.

Lemma 4.2 *Assume* $\psi(1) = 0$. *Let* $a \in D$. *If there exists a* σ-*field* $\mathscr{G} \subset \mathscr{F}$ *such that*

$$\liminf_{n \to \infty} \mathbf{Q}_a^{(h)}(M_n^{(h)} \mid \mathscr{G}) < \infty, \quad \mathbf{Q}_a^{(h)}\text{-a.s.}, \tag{4.7}$$

the $\mathbf{P}_a$-*martingale* $(M_n^{(h)}, n \geq 0)$ *is uniformly integrable. In particular,* $\mathbf{E}_a(M_\infty^{(h)}) = h(a)e^{-a}$.

Proof We claim that $\frac{1}{M_n^{(h)}}$ is a $\mathbf{Q}_a^{(h)}$-supermartingale (warning: it is a common mistake to claim that $\frac{1}{M_n^{(h)}}$ is a $\mathbf{Q}_a^{(h)}$-martingale; see discussions in Harris and Roberts [125]): Let $n \geq j$ and $A \in \mathscr{F}_j$; we have

$$\mathbf{Q}_a^{(h)}\left(\frac{1}{M_n^{(h)}} \mathbf{1}_A\right) = \mathbf{P}\{M_n^{(h)} > 0, A\} \leq \mathbf{P}\{M_j^{(h)} > 0, A\} = \mathbf{Q}_a^{(h)}\left(\frac{1}{M_j^{(h)}} \mathbf{1}_A\right),$$

which implies $\mathbf{Q}_a^{(h)}[\frac{1}{M_n^{(h)}} \mid \mathscr{F}_j] \leq \frac{1}{M_j^{(h)}}$, and proves the claimed supermartingale property.

Since this supermartingale is non-negative, there exists a (finite) random variable $L_\infty^{(h)} \geq 0$ such that $\frac{1}{M_n^{(h)}} \to L_\infty^{(h)}$, $\mathbf{Q}_a^{(h)}$-a.s.

If (4.7) is satisfied, then by the conditional Fatou's lemma,

$$\mathbf{Q}_a^{(h)}\left(\frac{1}{L_\infty^{(h)}} \mid \mathscr{G}\right) < \infty, \quad \mathbf{Q}_a^{(h)}\text{-a.s.}$$

In particular, $\frac{1}{L_\infty^{(h)}} < \infty$, $\mathbf{Q}_a^{(h)}$-a.s., so $\sup_{n \geq 0} M_n^{(h)} < \infty$, $\mathbf{Q}_a^{(h)}$-a.s.

For $u > 0$,

$$\mathbf{E}_a[M_n^{(h)} \mathbf{1}_{\{M_n^{(h)} > u\}}] = \mathbf{Q}_a^{(h)}\{M_n^{(h)} > u\} \leq \mathbf{Q}_a^{(h)}\left\{\sup_{n \geq 0} M_n^{(h)} > u\right\},$$

which tends to 0 as $u \to \infty$, uniformly in n. So $(M_n^{(h)}, n \geq 0)$ is uniformly integrable under $\mathbf{P}_a$. $\square$

4.4 The Spinal Decomposition Theorem

We assume that $\psi(1) = 0$, i.e., $\mathbf{E}[\sum_{|x|=1} e^{-V(x)}] = 1$. Let $D \subset \mathbb{R}$ be a Borel set satisfying (4.2), and let $a \in D$. Let $h : D \to (0, \infty)$ be a positive harmonic function in the sense of (4.3).

For any $b \in D$, let $\hat{\Xi}_b^{(h)} := (\hat{\xi}_i, 1 \leq i \leq \widehat{N})$ be such that for any sequence $(v_i, i \geq 1)$ of real numbers,

$$\mathbf{P}\left(\hat{\xi}_i \leq v_i, \ \forall 1 \leq i \leq \widehat{N}\right)$$

$$= \mathbf{E}\left[\mathbf{1}_{\{\xi_i + b \leq v_i, \ \forall 1 \leq i \leq N\}} \frac{\sum_{j=1}^N h(\xi_j + b)e^{-(\xi_j + b)} \mathbf{1}_{\{\xi_j + b \in D\}}}{h(b)e^{-b}}\right]. \quad (4.8)$$

It is immediately seen that $\widehat{N} \geq 1$ almost surely.

We introduce the following new system,[3] which is a branching random walk with a spine $w^{(h)} = (w_n^{(h)}, n \geq 0)$:

- Initially, there is one particle $w_0^{(h)} := \varnothing$ at position $V(w_0^{(h)}) = a$.
- At time 1, the particle $w_0^{(h)}$ dies, giving birth to new particles distributed as $\hat{\Xi}_{V(w_0^{(h)})}^{(h)}$; the particle $w_1^{(h)}$ is chosen among the children y of $w_0^{(h)}$ with probability proportional to $h(V(y))e^{-V(y)} \mathbf{1}_{\{h(V(y)) \in D\}}$, while all other particles (if any) are normal particles.
- More generally, at each time $n \geq 1$, all particles die, while giving birth independently to sets of new particles. The children of normal particles z are distributed as $\Xi_{V(z)}$. The children of the particle $w_{n-1}^{(h)}$ are distributed as $\hat{\Xi}_{V(w_{n-1}^{(h)})}^{(h)}$; the particle $w_n^{(h)}$ is chosen among the children y of $w_{n-1}^{(h)}$ with probability proportional to $h(V(y))e^{-V(y)} \mathbf{1}_{\{h(V(y)) \in D\}}$; all other particles (if any) in the n-th generation are normal.
- The system goes on indefinitely.

See Fig. 4.1.

---

[3]Notation: the law of Ξ_r is defined as the law of $(\xi_i + r, 1 \leq i \leq N)$.

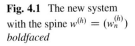

Fig. 4.1 The new system
with the spine $w^{(h)} = (w_n^{(h)})$
boldfaced

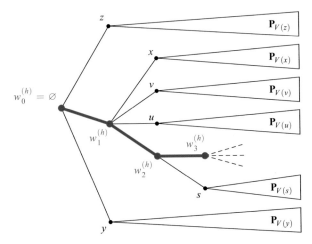

We note that while it is possible for a normal particle to produce no child if $P(N = 0) > 0$, the particles in the spine w are ensured to have at least one child because $\widehat{N} \geq 1$, a.s. Moreover, by the definition of $\widehat{\Xi}_b^{(h)}$, for any n, there exists at least a child y of $w_{n-1}^{(h)}$ such that $h(V(y))e^{-V(y)} \mathbf{1}_{\{V(y)\in D\}} > 0$, so there is almost surely no extinction of the new system.

Let us denote by $\mathscr{B}_a^{(h)}$ the law of the new system. It is a probability measure on the product space between the space of all marked trees (where the branching random walk lives), and the space of all infinite rays (where the spine $w^{(h)}$ lives), though we do not need to know anything particular about this product space. By an abuse of notation, the projection of $\mathscr{B}_a^{(h)}$ on the space of marked trees is still denoted by $\mathscr{B}_a^{(h)}$. The following theorem tells us that $\mathscr{B}_a^{(h)}$ describes precisely the law of the branching random walk $(V(x))$ under the probability $\mathbf{Q}_a^{(h)}$.

Theorem 4.3 (The Spinal Decomposition Theorem) *Assume $\psi(1) = 0$. Let $D \subset \mathbb{R}$ be a Borel set satisfying (4.2). For any $a \in D$, and any positive harmonic function h on D, the law of the branching random walk $(V(x))$ under $\mathbf{Q}_a^{(h)}$ is $\mathscr{B}_a^{(h)}$, where $\mathbf{Q}_a^{(h)}$ is the probability defined in (4.6).*

Along the spine $w^{(h)} = (w_n^{(h)})$, the probabilistic behaviour of the branching random walk under the new probability $\mathbf{Q}_a^{(h)}$ is particularly simple, as seen in the following theorem. Let, as before, (S_n) be an associated random walk in the sense of (4.1).

Theorem 4.4 (Along the Spine) *Assume $\psi(1) = 0$. Let $D \subset \mathbb{R}$ be a Borel set satisfying (4.2). Let $a \in D$, and let h be a positive harmonic function on D. Let $\mathbf{Q}_a^{(h)}$ be the probability defined in (4.6).*

(i) *For any $n \geq 0$ and any vertex $x \in \mathbb{T}$ with $|x| = n$,*

$$\mathbf{Q}_a^{(h)}(w_n^{(h)} = x \mid \mathscr{F}_n) = \frac{h(V(x))\mathrm{e}^{-V(x)}\,\mathbf{1}_{\{V(y)\in\mathrm{D},\,\forall y\in[\![\varnothing,\,x]\!]\}}}{M_n^{(h)}}, \tag{4.9}$$

where $M_n^{(h)} := \sum_{|z|=n} h(V(z))\mathrm{e}^{-V(z)}\,\mathbf{1}_{\{V(y)\in\mathrm{D},\,\forall y\in[\![\varnothing,\,z]\!]\}}$ as in (4.5).

(ii) *The process $(V(w_n^{(h)}),\ n \geq 0)$ under $\mathbf{Q}_a^{(h)}$ is distributed as the random walk $(S_n,\ n \geq 0)$ under $\mathbf{P}_a$ conditioned to stay in D (in the sense of (4.4)).*

We mention that since $\mathbf{Q}_a^{(h)}(M_n^{(h)} > 0) = 1$, the ratio on the right-hand side of (4.9) is $\mathbf{Q}_a^{(h)}$-almost surely well-defined.

Theorem 4.4(ii) has the following equivalent statement: For any $n \geq 1$ and any measurable function $g : \mathbb{R}^n \to [0, \infty)$,

$$\mathbf{E}_{\mathbf{Q}_a^{(h)}}[g(V(w_i^{(h)}),\ 0 \leq i \leq n)] = \mathbf{E}_a\left[g(S_i,\ 0 \leq i \leq n)\,\frac{h(S_n)}{h(a)}\,\mathbf{1}_{\{S_i\in\mathrm{D},\,\forall i\in[0,n]\cap\mathbb{Z}\}}\right]. \tag{4.10}$$

Theorems 4.3 and 4.4 are proved in the next section.

4.5   Proof of the Spinal Decomposition Theorem

This section is devoted to the proof of Theorems 4.3 and 4.4.

Proof of Theorem 4.3 We assume $\psi(1) = 0$ and fix D, a and h as stated in the theorem.

Following Neveu [203], we encode our genealogical tree $\mathbb{T}$ with $\mathscr{U} := \{\varnothing\} \cup \bigcup_{n=1}^{\infty}(\mathbb{N}^*)^n$. Let $(\phi_x,\ x \in \mathscr{U})$ be a family of non-negative Borel functions. If $\mathbf{E}_{\mathscr{B}_a^{(h)}}$ stands for expectation with respect to $\mathscr{B}_a^{(h)}$, we need to show that for any integer n,

$$\mathbf{E}_{\mathscr{B}_a^{(h)}}\left\{\prod_{|x|\leq n}\phi_x(V(x))\right\} = \mathbf{E}_{\mathbf{Q}_a^{(h)}}\left\{\prod_{|x|\leq n}\phi_x(V(x))\right\},$$

or, equivalently (by definition of $\mathbf{Q}_a^{(h)}$),

$$\mathbf{E}_{\mathscr{B}_a^{(h)}}\left\{\prod_{|x|\leq n}\phi_x(V(x))\right\} = \mathbf{E}_a\left\{\frac{M_n^{(h)}}{h(a)\mathrm{e}^{-a}}\prod_{|x|\leq n}\phi_x(V(x))\right\}. \tag{4.11}$$

For brevity, let us write

$$h_{\mathrm{D}}(x) := \frac{h(V(x))\mathrm{e}^{-V(x)}}{h(a)\mathrm{e}^{-a}}\,\mathbf{1}_{\{V(y)\in\mathrm{D},\,\forall y\in[\![\varnothing,\,x]\!]\}}\,\mathbf{1}_{\{x\in\mathbb{T}\}}.$$

If we are able to prove that for any $z \in \mathcal{U}$ with $|z| = n$,

$$\mathbf{E}_{\mathscr{B}_a^{(h)}} \left\{ \mathbf{1}_{\{w_n^{(h)} = z\}} \prod_{|x| \leq n} \phi_x(V(x)) \right\} = \mathbf{E}_a \left\{ h_D(z) \prod_{|x| \leq n} \phi_x(V(x)) \right\}, \tag{4.12}$$

then this will obviously yield (4.11) by summing over $|z| = n$.

So it remains to check (4.12). For $x \in \mathcal{U}$, let $\mathbb{T}_x$ be the subtree rooted at x, and $\text{brot}(x)$ the set of the brothers of x. A vertex y of $\mathbb{T}_x$ corresponds to the vertex xy of $\mathbb{T}$ where xy is the element of $\mathcal{U}$ obtained by concatenation of x and y. By the construction of $\mathscr{B}_a^{(h)}$, a branching random walk emanating from a vertex $y \notin (w_n^{(h)}, n \geq 0)$ has the same law as the original branching random walk under $\mathbf{P}$. By decomposing the product inside $\mathbf{E}_{\mathscr{B}_a^{(h)}}\{\cdots\}$ along the path $[\![\varnothing, z]\!]$, we observe that

$$\mathbf{E}_{\mathscr{B}_a^{(h)}} \left\{ \mathbf{1}_{\{w_n^{(h)} = z\}} \prod_{|x| \leq n} \phi_x(V(x)) \right\}$$

$$= \mathbf{E}_{\mathscr{B}_a^{(h)}} \left\{ \mathbf{1}_{\{w_n^{(h)} = z\}} \prod_{k=0}^{n} \phi_{z_k}(V(z_k)) \prod_{x \in \text{brot}(z_k)} \Phi_x(V(x)) \right\},$$

where z_k is the ancestor of z at generation k (with $z_n = z$), and for any $t \in \mathbb{R}$ and $x \in \mathcal{U}$, $\Phi_x(t) := \mathbf{E}\{\prod_{y \in \mathbb{T}_x} \phi_{xy}(t + V(y)) \mathbf{1}_{\{|y| \leq n - |x|\}}\}$. [The function Φ_x has nothing to do with the function Φ in Sect. 3.3.] Similarly,

$$\mathbf{E}_a \left\{ h_D(z) \prod_{|x| \leq n} \phi_x(V(x)) \right\} = \mathbf{E}_a \left\{ h_D(z) \prod_{k=0}^{n} \phi_{z_k}(V(z_k)) \prod_{x \in \text{brot}(z_k)} \Phi_x(V(x)) \right\}.$$

Therefore, the proof of (4.12) is reduced to showing the following: For any n and $|z| = n$, and any non-negative Borel functions $(\phi_{z_k}, \Phi_x)_{k,x}$,

$$\mathbf{E}_{\mathscr{B}_a^{(h)}} \left\{ \mathbf{1}_{\{w_n^{(h)} = z\}} \prod_{k=0}^{n} \phi_{z_k}(V(z_k)) \prod_{x \in \text{brot}(z_k)} \Phi_x(V(x)) \right\}$$

$$= \mathbf{E}_a \left\{ h_D(z) \prod_{k=0}^{n} \phi_{z_k}(V(z_k)) \prod_{x \in \text{brot}(z_k)} \Phi_x(V(x)) \right\}. \tag{4.13}$$

We prove (4.13) by induction. For $n = 0$, (4.13) is trivially true. Assume that the equality holds for $n - 1$ and let us prove it for n. By the definition of $\mathscr{B}_a^{(h)}$, given that $w_{n-1}^{(h)} = z_{n-1}$, the probability to choose $w_n^{(h)} = z$ among the children of $w_{n-1}^{(h)}$ is proportional to $h_D(z)$. Therefore, if we write

$$\Psi(z) := \phi_z(V(z)) \prod_{x \in \text{brot}(z)} \Phi_x(V(x)),$$

$$\mathscr{G}_{n-1}^{(h)} := \sigma\{w_k^{(h)}, V(w_k^{(h)}), \text{brot}(w_k^{(h)}), (V(y))_{y \in \text{brot}(w_k^{(h)})}, 1 \leq k \leq n-1\},$$

then

$$
\mathbf{E}_{\mathscr{B}_a^{(h)}} \left\{ \mathbf{1}_{\{w_n^{(h)} = z\}} \, \Psi(z) \,\middle|\, \mathscr{G}_{n-1}^{(h)} \right\}
$$

$$
= \mathbf{1}_{\{w_{n-1}^{(h)} = z_{n-1}\}} \, \mathbf{E}_{\mathscr{B}_a^{(h)}} \left\{ \frac{h_{\mathrm{D}}(z)}{h_{\mathrm{D}}(z) + \sum_{x \in \mathrm{brot}(z)} h_{\mathrm{D}}(x)} \, \Psi(z) \,\middle|\, \mathscr{G}_{n-1}^{(h)} \right\}
$$

$$
= \mathbf{1}_{\{w_{n-1}^{(h)} = z_{n-1}\}} \, \mathbf{E}_{\mathscr{B}_a^{(h)}} \left\{ \frac{h_{\mathrm{D}}(z)}{h_{\mathrm{D}}(z) + \sum_{x \in \mathrm{brot}(z)} h_{\mathrm{D}}(x)} \, \Psi(z) \,\middle|\, (w_{n-1}^{(h)}, \, V(w_{n-1}^{(h)})) \right\}.
$$

By assumption, the point process generated by $w_{n-1}^{(h)} = z_{n-1}$ has Radon–Nikodym derivative $\frac{h_{\mathrm{D}}(z) + \sum_{x \in \mathrm{brot}(z)} h_{\mathrm{D}}(x)}{h_{\mathrm{D}}(z_{n-1})}$ with respect to the point process generated by z_{n-1} under $\mathbf{P}_a$. Thus, on $\{w_{n-1}^{(h)} = z_{n-1}\}$,

$$
\mathbf{E}_{\mathscr{B}_a^{(h)}} \left\{ \frac{h_{\mathrm{D}}(z)}{h_{\mathrm{D}}(z) + \sum_{x \in \mathrm{brot}(z)} h_{\mathrm{D}}(x)} \, \Psi(z) \,\middle|\, (w_{n-1}^{(h)}, \, V(w_{n-1}^{(h)})) \right\}
$$

$$
= \mathbf{E}_a \left\{ \frac{h_{\mathrm{D}}(z)}{h_{\mathrm{D}}(z_{n-1})} \, \Psi(z) \,\middle|\, V(z_{n-1}) \right\}
$$

$$
=: \mathrm{RHS}_{(4.14)}. \tag{4.14}
$$

This implies

$$
\mathbf{E}_{\mathscr{B}_a^{(h)}} \left\{ \mathbf{1}_{\{w_n^{(h)} = z\}} \, \Psi(z) \,\middle|\, \mathscr{G}_{n-1}^{(h)} \right\} = \mathbf{1}_{\{w_{n-1}^{(h)} = z_{n-1}\}} \, \mathrm{RHS}_{(4.14)}.
$$

Let $\mathrm{LHS}_{(4.13)}$ denote the expression on the left-hand side of (4.13). Then

$$
\mathrm{LHS}_{(4.13)} = \mathbf{E}_{\mathscr{B}_a^{(h)}} \left\{ \mathbf{1}_{\{w_n^{(h)} = z\}} \, \Psi(z) \prod_{k=0}^{n-1} \phi_{z_k}(V(z_k)) \prod_{x \in \mathrm{brot}(z_k)} \Phi_x(V(x)) \right\}
$$

$$
= \mathbf{E}_{\mathscr{B}_a^{(h)}} \left\{ \mathbf{1}_{\{w_{n-1}^{(h)} = z_{n-1}\}} \, \mathrm{RHS}_{(4.14)} \prod_{k=0}^{n-1} \phi_{z_k}(V(z_k)) \prod_{x \in \mathrm{brot}(z_k)} \Phi_x(V(x)) \right\},
$$

which, by the induction hypothesis, is

$$
= \mathbf{E}_a \left\{ h_{\mathrm{D}}(z_{n-1}) \, \mathrm{RHS}_{(4.14)} \prod_{k=0}^{n-1} \phi_{z_k}(V(z_k)) \prod_{x \in \mathrm{brot}(z_k)} \Phi_x(V(x)) \right\}
$$

$$
= \mathbf{E}_a \left\{ h_{\mathrm{D}}(z) \prod_{k=0}^{n} \phi_{z_k}(V(z_k)) \prod_{x \in \mathrm{brot}(z_k)} \Phi_x(V(x)) \right\},
$$

where the last equality follows from the fact that $\mathrm{RHS}_{(4.14)}$ is also the conditional $\mathbf{E}_a$-expectation of $\frac{h_\mathrm{D}(z)}{h_\mathrm{D}(z_{n-1})}\phi_z(V(z))\prod_{x\in\mathrm{brot}(z)}\Phi_x(V(x))$ given $V(z_k)$, $\mathrm{brot}(z_k)$ and $(V(x),\ x\in\mathrm{brot}(z_k))$, $0\le k\le n-1$. This yields (4.13), and completes the proof of Theorem 4.3. $\qquad\square$

Proof of Theorem 4.4 We assume $\psi(1)=0$ and fix D, a and h as stated in the theorem.

Let $(\phi_x, x\in\mathscr{U})$ be a family of Borel functions and $z\in\mathscr{U}$ a vertex with $|z|=n$. Write again $h_\mathrm{D}(x) := \frac{h(V(x))e^{-V(x)}}{h(a)e^{-a}}\mathbf{1}_{\{V(y)\in\mathrm{D},\ \forall y\in[\varnothing,x]\}}$. By (4.12) (identifying $\mathbf{E}_{\mathscr{B}_a^{(h)}}$ with $\mathbf{E}_{\mathbf{Q}_a^{(h)}}$ by Theorem 4.3),

$$\mathbf{E}_{\mathbf{Q}_a^{(h)}}\Big\{\mathbf{1}_{\{w_n^{(h)}=z\}}\prod_{|x|\le n}\phi_x(V(x))\Big\} = \mathbf{E}_a\Big\{h_\mathrm{D}(z)\prod_{|x|\le n}\phi_x(V(x))\Big\},$$

which, by the definition of $\mathbf{Q}_a^{(h)}$, is

$$= \mathbf{E}_{\mathbf{Q}_a^{(h)}}\Big\{\frac{h(a)e^{-a}h_\mathrm{D}(z)}{M_n^{(h)}}\prod_{|x|\le n}\phi_x(V(x))\Big\},$$

where $M_n^{(h)}$ is the martingale defined in (4.5). This shows that

$$\mathbf{Q}_a^{(h)}(w_n^{(h)}=z\mid\mathscr{F}_n) = \frac{h(a)e^{-a}h_\mathrm{D}(z)}{M_n^{(h)}}, \qquad (4.15)$$

proving part (i) of the theorem.

To prove part (ii), we take $n\ge 1$ and a measurable function $g:\mathbb{R}^{n+1}\to[0,\infty)$. Write $\mathbf{E}_{\mathbf{Q}_a^{(h)}}[g]$ for $\mathbf{E}_{\mathbf{Q}_a^{(h)}}[g(V(w_i^{(h)}),\ 0\le i\le n)]$. Then

$$\mathbf{E}_{\mathbf{Q}_a^{(h)}}[g] = \mathbf{E}_{\mathbf{Q}_a^{(h)}}\Big(\sum_{|x|=n}g(V(x_i),\ 0\le i\le n)\,\mathbf{1}_{\{w_n^{(h)}=x\}}\Big).$$

By (4.15), we get

$$\mathbf{E}_{\mathbf{Q}_a^{(h)}}[g] = \mathbf{E}_{\mathbf{Q}_a^{(h)}}\Big(\sum_{|x|=n}g(V(x_i),\ 0\le i\le n)\,\frac{h(a)e^{-a}h_\mathrm{D}(x)}{M_n^{(h)}}\Big)$$

$$= \mathbf{E}_a\Big(\sum_{|x|=n}g(V(x_i),\ 0\le i\le n)\,h_\mathrm{D}(x)\Big),$$

the last identity being a consequence of the definition of $\mathbf{Q}_a^{(h)}$. Plugging in the definition of h_D yields

$$\mathbf{E}_{\mathbf{Q}_a^{(h)}}[g] = \mathbf{E}_a\left(\sum_{|x|=n} g(V(x_i),\ 0 \le i \le n)\frac{h(V(x))\mathrm{e}^{-V(x)}}{h(a)\mathrm{e}^{-a}}\mathbf{1}_{\{V(y)\in D,\ \forall y\in[\varnothing,x]\}}\right)$$

$$= \mathbf{E}_a\left(g(S_i,\ 0 \le i \le n)\frac{h(S_n)}{h(a)}\mathbf{1}_{\{S_i\in D,\ \forall i\in[0,n]\cap\mathbb{Z}\}}\right), \quad \text{(many-to-one)}$$

yielding part (ii) of Theorem 4.4. $\qquad\square$

4.6 Example: Size-Biased Branching Random Walks

We present an important example of the spinal decomposition for branching random walks.

Example 4.5 (Size-Biased Branching Random Walks) Assume $\psi(1) = 0$. We take $D = \mathbb{R}$, so that (4.2) is automatically satisfied. Let $h(u) := 1,\ u \in \mathbb{R}$, which is trivially harmonic. The martingale $M_n^{(h)}$ defined in (4.5) is nothing else but the additive martingale[4]

$$M_n = W_n = \sum_{|x|=n} \mathrm{e}^{-V(x)}.$$

The change of probabilities in (4.6) becomes: for $a \in \mathbb{R}$,

$$\mathbf{Q}_a(A) = \int_A \frac{W_n}{\mathrm{e}^{-a}}\,\mathrm{d}\mathbf{P}_a, \quad \forall A \in \mathscr{F}_n,\ \forall n \ge 0.$$

The new point process $\hat{\Xi}_b$ (for $b \in \mathbb{R}$) in (4.8) becomes: For any sequence $(v_i,\ i \ge 1)$ of real numbers,

$$\mathbf{P}\left(\hat{\xi}_i \le v_i,\ \forall 1 \le i \le \widehat{N}\right) = \mathbf{E}\left[\mathbf{1}_{\{\xi_i+b\le v_i,\ \forall 1\le i\le N\}}\sum_{j=1}^N \mathrm{e}^{-\xi_j}\right].$$

---

[4]Since the harmonic function is a constant in Example 4.5, we drop the superscript h in $M_n^{(h)}$, $\mathbf{Q}_a^{(h)}$, $\hat{\Xi}_b^{(h)}$, $w^{(h)}$, etc.

The spinal decomposition theorem tells us that for any $a \in \mathbb{R}$, under $\mathbf{Q}_a$, the branching random walk can be provided with a spine (w_n) such that:

- Each spine particle w_n gives birth to a set of new particles according to the distribution of $\hat{\Xi}_{V(w_n)}$; we choose the particle w_{n+1} among the children y of w_n with probability proportional to $e^{-V(y)}$; all other particles are normal particles.
- Children of a normal particle z are normal, and are distributed as $\Xi_{V(z)}$.

The law of the branching random walk under $\mathbf{Q}_a$ is often referred to as the law of the **size-biased branching random walk**. It is clear that if $V(x) = 0$ for all x, then the description of the law of the size-biased branching random walk coincides with the description of the law of the size-biased Galton–Watson tree in Sect. 2.2.

By Theorem 4.4 in Sect. 4.4,

$$\mathbf{Q}_a(w_n = x \mid \mathscr{F}_n) = \frac{e^{-V(x)}}{W_n}, \tag{4.16}$$

for any n and any vertex x such that $|x| = n$, and under $\mathbf{Q}_a$, $(V(w_n) - V(w_{n-1})$), $n \geq 1$) is a sequence of i.i.d. random variables whose common distribution is that of S_1 under $\mathbf{P}_0$. $\square$

4.7   Example: Above a Given Value Along the Spine

We now give another important example of the spinal decomposition theorem.

Example 4.6 (Branching Random Walk Above a Given Value Along the Spine) Assume $\psi(1) = 0$. Fix $\alpha \geq 0$.

Let (S_n) be an associated random walk in the sense of (4.1). Let R be the renewal function associated with (S_n), as in Appendix A.1. Define the function $h : [-\alpha, \infty) \to (0, \infty)$ by

$$h(u) := R(u + \alpha), \quad u \in [-\alpha, \infty). \tag{4.17}$$

Lemma 4.7 below says that h is a positive harmonic function on $D := [-\alpha, \infty)$ in the sense of (4.3), so the spine decomposition theorem applies in this situation. Along the spine, the process $(V(w_n^{(h)}), n \geq 0)$ under $\mathbf{Q}_a^{(h)}$ is distributed as the random walk $(S_n, n \geq 0)$ under $\mathbf{P}_a$ conditioned to be $\geq -\alpha$ (in the sense of (4.4)).

We mention that even though the branching random walk stays above $-\alpha$ along the spine, it can certainly hit $(-\infty, -\alpha)$ off the spine.

Let us complete the presentation of Example 4.6 with the following lemma.

Lemma 4.7 *Let $\alpha \geq 0$ and let h be as in (4.17). Then*

$$h(a) = \mathbf{E}_a[h(S_1) \, \mathbf{1}_{\{S_1 \geq -\alpha\}}], \quad \forall a \geq -\alpha.$$

Proof It boils down to checking that

$$R(b) = \mathbf{E}[R(S_1 + b) \mathbf{1}_{\{S_1 \geq -b\}}], \quad \forall b \geq 0. \tag{4.18}$$

Let $\tau^+ := \inf\{k \geq 1 : S_k \geq 0\}$, which is well-defined almost surely, since $\mathbf{E}(S_1) = 0$. If $\widetilde{S}_1$ is a random variable independent of $(S_i, i \geq 1)$, then

$$\mathbf{E}[R(S_1 + b) \mathbf{1}_{\{S_1 \geq -b\}}] = \mathbf{E}\Big[\sum_{j=0}^{\tau^+ - 1} \mathbf{1}_{\{S_j \geq -\widetilde{S}_1 - b, \, \widetilde{S}_1 \geq -b\}}\Big]$$

$$= \mathbf{E}\Big[\sum_{j=0}^{\infty} \mathbf{1}_{\{\tau^+ > j\}} \mathbf{1}_{\{S_j \geq -\widetilde{S}_1 - b, \, \widetilde{S}_1 \geq -b\}}\Big].$$

On the event $\{\tau^+ > j\} \cap \{S_j \geq -\widetilde{S}_1 - b\}$, we automatically have $\widetilde{S}_1 \geq -b$ (because $S_j \geq -\widetilde{S}_1 - b$ while $S_j < 0$). Therefore,

$$\mathbf{E}[R(S_1 + b) \mathbf{1}_{\{S_1 \geq -b\}}] = \mathbf{E}\Big[\sum_{j=0}^{\infty} \mathbf{1}_{\{S_j \geq -\widetilde{S}_1 - b, \, S_i < 0, \, \forall i \leq j\}}\Big]$$

$$= \sum_{j=0}^{\infty} \mathbf{P}\{S_j \geq -\widetilde{S}_1 - b, \, S_i < 0, \, \forall i \leq j\}.$$

For any j, $\mathbf{P}\{S_j \geq -\widetilde{S}_1 - b, \, S_i < 0, \, \forall i \leq j\} = \mathbf{P}\{S_{j+1} \geq -b, \, S_i < 0, \, \forall i \leq j\}$. By splitting $\{S_{j+1} \geq -b\}$ as $\{S_{j+1} \geq 0\} \cup \{-b \leq S_{j+1} < 0\}$, this leads to:

$$\mathbf{E}[R(S_1 + b) \mathbf{1}_{\{S_1 \geq -b\}}] = \sum_{j=0}^{\infty} \mathbf{P}\{S_{j+1} \geq 0, \, S_i < 0, \, \forall i \leq j\}$$

$$+ \sum_{j=0}^{\infty} \mathbf{P}\{-b \leq S_{j+1} < 0, \, S_i < 0, \, \forall i \leq j\}.$$

The first sum on the right-hand side is equal to 1: it suffices to note that $\{S_{j+1} \geq 0, \, S_i < 0, \, \forall i \leq j\} = \{\tau^+ = j + 1\}$ and that $\sum_{j=0}^{\infty} \mathbf{P}\{\tau^+ = j + 1\} = 1$. Therefore,

$$\mathbf{E}[R(S_1 + b) \mathbf{1}_{\{S_1 \geq -b\}}] = 1 + \sum_{j=0}^{\infty} \mathbf{P}\{-b \leq S_{j+1} < 0, \, S_i < 0, \, \forall i \leq j\}. \tag{4.19}$$

On the other hand, by definition,

$$R(b) = \mathbf{E}\Big[\sum_{j=0}^{\infty} \mathbf{1}_{\{\tau^+ > j, \, S_j \geq -b\}}\Big],$$

which yields that

$$R(b) = 1 + \sum_{j=1}^{\infty} \mathbf{P}\{\tau^+ > j, \, S_j \geq -b\} = 1 + \sum_{j=0}^{\infty} \mathbf{P}\{\tau^+ > j+1, \, S_{j+1} \geq -b\}.$$

Since $\{\tau^+ > j+1, \, S_{j+1} \geq -b\} = \{-b \leq S_{j+1} < 0, \, S_i < 0, \, \forall i \leq j\}$, this yields $R(b) = 1 + \sum_{j=0}^{\infty} \mathbf{P}\{-b \leq S_{j+1} < 0, \, S_i < 0, \, \forall i \leq j\}$, which, in view of (4.19), is equal to $\mathbf{E}[R(S_1 + b) \mathbf{1}_{\{S_1 \geq -b\}}]$. This proves (4.18).   □

4.8   Application: The Biggins Martingale Convergence Theorem

We apply the spinal decomposition theorem to prove the Biggins martingale convergence theorem (Theorem 3.2 in Sect. 3.2), of which we recall the statement: Assume $\psi(0) > 0$, $\psi(1) < \infty$ and $\psi'(1) \in \mathbb{R}$. Let $W_n := \sum_{|x|=n} e^{-n\psi(1) - V(x)}$, and let W_∞ be the a.s. limit of W_n. Then

$$\mathbf{E}(W_\infty) = 1 \; \Leftrightarrow \; W_\infty > 0, \; \mathbf{P}^*\text{-a.s.}$$
$$\Leftrightarrow \; \mathbf{E}(W_1 \ln_+ W_1) < \infty \; \text{ and } \; \psi(1) > \psi'(1). \tag{4.20}$$

Proof of the Biggins Martingale Convergence Theorem Let $\mathbf{Q}$ be the probability on $\mathscr{F}_\infty$ such that $\mathbf{Q}_{|\mathscr{F}_n} = W_n \bullet \mathbf{P}_{|\mathscr{F}_n}$ for all $n \geq 0$.

(i) Assume that the last condition in (4.20) fails.

We claim that in this case, $\limsup_{n\to\infty} W_n = \infty$ $\mathbf{Q}$-a.s.; thus by (2.3) of Sect. 2.3, $\mathbf{E}(W_\infty) = 0$.

To prove our claim, let us distinguish two possibilities.

First possibility: $\psi(1) \leq \psi'(1)$. Under $\mathbf{Q}$, $(V(w_n) - V(w_{n-1}), n \geq 1)$ are i.i.d. (see Example 4.5 in Sect. 4.6), so by the law of large numbers, when $n \to \infty$,

$$\frac{V(w_n)}{n} \to \mathbf{E}_{\mathbf{Q}}[V(w_1)] = \mathbf{E}\Big[\sum_{|x|=1} V(x)e^{-V(x)-\psi(1)}\Big] = -\psi'(1), \quad \mathbf{Q}\text{-a.s.}$$

In other words, $\frac{-V(w_n)-n\psi(1)}{n} \to \psi'(1) - \psi(1)$, $\mathbf{Q}$-a.s.

If $\psi'(1) > \psi(1)$, this yields $-V(w_n) - n\psi(1) \to \infty$, $\mathbf{Q}$-a.s. If $\psi'(1) = \psi(1)$, the associated random walk $(-V(w_n) - n\psi(1), n \geq 1)$ being oscillating (under $\mathbf{Q}$), we have $\limsup_{n\to\infty}[-V(w_n) - n\psi(1)] = \infty$, $\mathbf{Q}$-a.s.

So as long as $\psi(1) \leq \psi'(1)$, we have $\limsup_{n\to\infty}[-V(w_n) - n\psi(1)] = \infty$, $\mathbf{Q}$-a.s. Since $W_n \geq e^{-V(w_n)-n\psi(1)}$, we get $\limsup_{n\to\infty} W_n = \infty$ $\mathbf{Q}$-a.s., as claimed.

Second possibility: $\mathbf{E}(W_1 \ln_+ W_1) = \infty$. In this case, we argue that (recalling that $\overleftarrow{y}$ is the parent of y)

$$W_{n+1} = \sum_{|x|=n} e^{-V(x)-(n+1)\psi(1)} \sum_{|y|=n+1: \ \overleftarrow{y}=x} e^{-[V(y)-V(x)]}$$

$$\geq e^{-V(w_n)-(n+1)\psi(1)} \sum_{|y|=n+1: \ \overleftarrow{y}=w_n} e^{-[V(y)-V(w_n)]}.$$

Write

$$W(w_n) := \sum_{|y|=n+1: \ \overleftarrow{y}=w_n} e^{-[V(y)-V(w_n)]}. \tag{4.21}$$

Since $W(w_n), n \geq 0$, are i.i.d. under $\mathbf{Q}$, with $\mathbf{E}_{\mathbf{Q}}[\ln_+ W(w_0)] = \mathbf{E}[W_1 \ln_+ W_1] = \infty$, it is a simple consequence of the Borel–Cantelli lemma that

$$\limsup_{n\to\infty} \frac{\ln W(w_n)}{n} = \infty, \quad \mathbf{Q}\text{-a.s.}$$

On the other hand, $\frac{V(w_n)}{n} \to -\psi'(1)$, $\mathbf{Q}$-a.s. (law of large numbers), this yields again $\limsup_{n\to\infty} W_n = \infty$ $\mathbf{Q}$-a.s., as claimed.

(ii) We now assume that the condition on the right-hand side of (4.20) is satisfied, i.e., $\psi'(1) < \psi(1)$ and $\mathbf{E}[W_1 \ln_+ W_1] < \infty$. Let $\mathscr{G}$ be the σ-field generated by w_n and $V(w_n)$ as well as the offspring of w_n, for all $n \geq 0$. Then $\mathbf{E}_{\mathbf{Q}}(W_n \mid \mathscr{G})$ is

$$= e^{-V(w_n)-n\psi(1)} + \sum_{k=0}^{n-1} e^{-V(w_k)-n\psi(1)} \sum_{|x|=k+1: \ \overleftarrow{x}=w_k, \ x\neq w_{k+1}} e^{-[V(x)-V(w_k)]} e^{[n-(k+1)]\psi(1)}$$

$$= \sum_{k=0}^{n-1} e^{-V(w_k)-(k+1)\psi(1)} \sum_{|x|=k+1: \ \overleftarrow{x}=w_k} e^{-[V(x)-V(w_k)]} - \sum_{i=1}^{n-1} e^{-V(w_i)-i\psi(1)}.$$

With our notation in (4.21), this reads as:

$$\mathbf{E}_{\mathbf{Q}}(W_n \mid \mathscr{G}) = \sum_{k=0}^{n-1} e^{-V(w_k)-(k+1)\psi(1)} W(w_k) - \sum_{k=1}^{n-1} e^{-V(w_k)-k\psi(1)}.$$

Since $W(w_n), n \geq 0$, are i.i.d. under $\mathbf{Q}$ with $\mathbf{E}_{\mathbf{Q}}[\ln_+ W(w_0)] = \mathbf{E}(W_1 \ln_+ W_1) < \infty$, it follows from the strong law of large numbers that

$$\frac{\ln_+ W(w_n)}{n} = \frac{1}{n}\sum_{i=1}^{n} \ln_+ W(w_i) - \frac{1}{n}\sum_{i=1}^{n-1} \ln_+ W(w_i) \to 0, \quad \mathbf{Q}\text{-a.s.}$$

On the other hand, $e^{-V(w_k)-(k+1)\psi(1)}$ and $e^{-V(w_k)-k\psi(1)}$ decay exponentially fast (because $\frac{-V(w_k)}{k} \to \psi'(1) < \psi(1)$, $\mathbf{Q}$-a.s., as $k \to \infty$). It follows that $\mathbf{E}_{\mathbf{Q}}[W_n \mid \mathscr{G}]$ converges $\mathbf{Q}$-a.s. (to a finite limit). By Lemma 4.2 (Sect. 4.3), this yields $\mathbf{E}(W_\infty) = 1$, which obviously implies $\mathbf{P}(W_\infty = 0) < 1$. Since we already know that $\mathbf{P}(W_\infty = 0)$ is either 1 or q, this implies that $\mathbf{P}(W_\infty = 0) = q$: the Biggins martingale convergence theorem is proved. □

4.9 Notes

The change of probabilities technique used in the spinal decomposition theorem (Sects. 4.3 and 4.4) has a long history, and has been employed by many people in various contexts. It goes back at least to Kahane and Peyrière [152].

A version of the spinal decomposition theorem associated with a function h which is not necessary harmonic can be found in Aïdékon et al. [17]. It is also possible to allow the harmonic function h to change from a certain generation on; see Addario-Berry et al. [4] for such an example for the Galton–Watson process.

The law of the size-biased branching random walk described in Example 4.5 (Sect. 4.6) is due to Lyons [173]. See also Waymire and Williams [232] for a formulation in a different language. The analogue for branching Brownian motion can be found in Chauvin and Rouault [85]; for more general diffusions, see Liu et al. [170].

The spinal decomposition for the branching random walk above a given level along the spine in Example 4.6 (Sect. 4.7) is borrowed from Biggins and Kyprianou [57]. Its analogue for branching Brownian motion is in Kyprianou [162]. The proof of Lemma 4.7 is due to Tanaka [226].

An extension of the spinal decomposition theorem for multiple spines, often convenient to compute higher-order moments, is proved by Harris and Roberts [126].

The short proof of the Biggins martingale convergence theorem in Sect. 4.8, via size-biased branching random walks, follows from Lyons [173].

Chapter 5
Applications of the Spinal Decomposition Theorem

Armed with the spinal decomposition theorem, we are now able to establish, in this chapter, some deep results for extreme values in the branching random walk. Among these results, a particularly spectacular one is the Aïdékon theorem for the limit distribution of the leftmost position (Sect. 5.4). We give a complete proof of the Aïdékon theorem by means of the peeling lemma (Theorem 5.14 in Sect. 5.3) for the spine, a very useful tool which is exploited in a few other situations in this chapter. Most of the applications of the spinal decomposition theorem presented here have been obtained in recent years, highlighting the state of the art of the study of branching random walks.

5.1   Assumption (H)

Let $(V(x), x \in \mathbb{T})$ be a branching random walk whose law is governed by a point process $\Xi := (\xi_1, \ldots, \xi_N)$, where the random variable N can be 0. Let ψ be the log-Laplace transform defined by

$$\psi(t) := \ln \mathbf{E}\left(\sum_{|x|=1} e^{-tV(x)} \right) \in (-\infty, \infty], \quad t \in \mathbb{R}.$$

Throughout this chapter, we assume $\psi(0) > 0$ and $\psi(1) = 0 = \psi'(1)$.

While $\psi(0) > 0$ clearly indicates that we work on the supercritical regime, let us say a few words about the assumption $\psi(1) = 0 = \psi'(1)$. Assume for the moment that there exists $t > 0$ such that $\psi(t) < \infty$. Let $\zeta := \sup\{s > 0 : \psi(s) < \infty\} \in (0, \infty]$. If we can find $t^* \in (0, \zeta)$ satisfying

$$\psi(t^*) = t^* \psi'(t^*), \tag{5.1}$$

© Springer International Publishing Switzerland 2015
Z. Shi, *Branching Random Walks*, Lecture Notes in Mathematics 2151,
DOI 10.1007/978-3-319-25372-5_5

then the log-Laplace transform of the branching random walk $\widetilde{V}(x) := t^* V(x) + \psi(t^*) |x|$, $x \in \mathbb{T}$, satisfies $\psi(1) = 0 = \psi'(1)$.

So as long as (5.1) has a solution, there is no loss of generality to assume that $\psi(1) = 0 = \psi'(1)$. It is, however, possible that no $t^* > 0$ satisfies (5.1), in which case, the results in this chapter do not apply. Loosely speaking, the existence of t^* fails if and only if the law of $\inf_i \xi_i$ is bounded from below and $\mathbf{E}(\sum_i \mathbf{1}_{\{\xi_i = \text{supp}_{\min}\}}) \geq 1$, with $\text{supp}_{\min}$ denoting the minimum of the support of the law of $\inf_i \xi_i$ (i.e., the essential infimum of $\inf_i \xi_i$). In particular, for a branching random walk with Gaussian displacements, t^* exists. For an elementary but complete discussion on the existence of solutions to (5.1) under the assumption $\psi(0) < \infty$, see the arXiv version of Jaffuel [146], or Bérard and Gouéré [41] assuming that $\#\varXi$ is bounded.

If $t > 0$ is such that $\psi(t) = 0$, the value of t is often referred to as a "Malthusian parameter" in various contexts ([47, 147, 202], etc.), because it governs the growth rate (in the exponential scale) of the system. Under our assumption, 1 is the unique Malthusian parameter.

The assumption $\psi(1) = 0 = \psi'(1)$ is fundamental for various universality behaviours of the branching random walk we are going to study in this chapter.

Recall that the one-dimensional random walk associated with $(V(x), x \in \mathbb{T})$, denoted by $(S_n, n \geq 0)$, is such that for any Borel function $g : \mathbb{R} \to [0, \infty)$,

$$\mathbf{E}[g(S_1)] = \mathbf{E}\left[\sum_{|x|=1} g(V(x)) e^{-V(x)} \right].$$

For the majority of the results in the chapter, we also assume

$$\mathbf{E}\left[\sum_{|x|=1} V(x)^2 e^{-V(x)} \right] < \infty. \tag{5.2}$$

Condition (5.2) simply says that $\mathbf{E}[S_1^2] < \infty$. Moreover, the assumption $\psi'(1) = 0$ means that $\mathbf{E}[S_1] = 0$.

Finally, we often also assume that

$$\mathbf{E}[X \ln_+^2 X] < \infty, \qquad \mathbf{E}[\widetilde{X} \ln_+ \widetilde{X}] < \infty, \tag{5.3}$$

where $\ln_+ y := \max\{0, \ln y\}$ and $\ln_+^2 y := (\ln_+ y)^2$ for any $y \geq 0$, and[1]

$$X := \sum_{|x|=1} e^{-V(x)}, \qquad \widetilde{X} := \sum_{|x|=1} V(x)_+ e^{-V(x)}. \tag{5.4}$$

---

[1] Notation: $u_+ := \max\{u, 0\}$ for any $u \in \mathbb{R}$.

Assumption (5.3), despite its somehow exotic looking, is believed to be optimal for most of the results in this chapter (for example, for the derivative martingale to have a positive limit; see Conjecture 5.9 in Sect. 5.2 for more details).

Summarizing, here is the fundamental assumption of the chapter:

Assumption (H) $\psi(0) > 0$, $\psi(1) = 0 = \psi'(1)$, (5.2) and (5.3).

The following elementary lemma tells us that under assumption (5.3), we have

$$\mathbf{E}[X \ln^2_+ \widetilde{X}] < \infty, \qquad \mathbf{E}[\widetilde{X} \ln_+ X] < \infty. \qquad (5.5)$$

Lemma 5.1 *Let ξ and η be an arbitrary pair of non-negative random variables such that $\mathbf{E}[\xi \ln^2_+ \xi] < \infty$ and that $\mathbf{E}[\eta \ln_+ \eta] < \infty$. Then*

$$\mathbf{E}[\xi \ln^2_+ \eta] < \infty, \qquad \mathbf{E}[\eta \ln_+ \xi] < \infty.$$

Proof The second inequality follows from the simple observation that $\eta \ln_+ \xi \le \max\{\xi \ln_+ \xi, \eta \ln_+ \eta\}$, whereas the first from the existence of a constant $c_1 > 0$ such that for all sufficiently large a and b,

$$a \ln^2 b \le c_1 (a \ln^2 a + b \ln b),$$

which is easily seen to hold by discussing on whether $b \le a^2$ or $b > a^2$. □

5.2 Convergence of the Derivative Martingale

In this section, we study the derivative martingale defined by

$$D_n := \sum_{|x|=n} V(x) e^{-V(x)}, \quad n \ge 0.$$

If $\psi(1) = 0 = \psi'(1)$, then (D_n) is indeed a martingale. Recall that $\mathbf{P}^*(\cdot) := \mathbf{P}(\cdot \,|\, \text{non-extinction})$.

Theorem 5.2

(i) *If $\psi(0) > 0$, $\psi(1) = 0 = \psi'(1)$ and [2] $\mathbf{E}[\sum_{|x|=1} V(x)^2_- e^{-V(x)}] < \infty$, then $(D_n, n \ge 0)$ converges a.s. to a non-negative limit, denoted by D_∞.*

(ii) *Under Assumption (H), we have $D_\infty > 0$, $\mathbf{P}^*$-a.s.*

Discussion 5.3 Even though the martingale D_n can be negative, we trivially see that its almost sure limit D_∞ is necessarily non-negative: indeed, since $\psi(1) = 0$,

---

[2]Notation: $u_- := \max\{-u, 0\}$ for any $u \in \mathbb{R}$.

Lemma 3.1 of Sect. 3.1 tells us that $\inf_{|x|=n} V(x) \to \infty$, $\mathbf{P}^*$-a.s., hence $D_n \geq 0$ $\mathbf{P}^*$-almost surely for all sufficiently large index n (how large depends on the underlying ω). So Theorem 5.2(i) reveals the almost sure convergence of (D_n).   □

Discussion 5.4 Let us now turn to Theorem 5.2 (ii). Assume for the moment that $\psi(0) > 0$, $\psi(1) = 0 = \psi'(1)$ and $\mathbf{E}[\sum_{|x|=1} V(x)^2 \, e^{-V(x)}] < \infty$. The first part of the theorem ensures the almost sure existence of D_∞. For any $x \in \mathbb{T}$, we write $D_{n,x} := \sum_{|y|=n:\, y \geq x}[V(y) - V(x)] e^{-[V(y)-V(x)]}$; then

$$D_{\infty,x} := \lim_{n \to \infty} D_{n,x} \tag{5.6}$$

exists a.s. as well, and is a.s. non-negative. For any $x \in \mathbb{T}$ with $|x| = 1$, we have

$$D_n \geq \sum_{|y|=n,\, y \geq x} V(y) e^{-V(y)} = e^{-V(x)} D_{n,x} + V(x) e^{-V(x)} \sum_{|y|=n,\, y \geq x} e^{-[V(y)-V(x)]}.$$

We let $n \to \infty$. Since $\psi(1) = 0 = \psi'(1)$, it follows from the Biggins martingale convergence theorem (Theorem 3.2 in Sect. 3.2) that $\sum_{|y|=n,\, y \geq x} e^{-[V(y)-V(x)]} \to 0$ a.s., so that

$$D_\infty \geq e^{-V(x)} D_{\infty,x}.$$

Since both D_∞ and $D_{\infty,x}$ are non-negative, this implies that $\{D_\infty = 0\} \subset \{D_{\infty,x} = 0\}$. In the literature, a property is called *inherited* [208, Chap. 3], [175, Chap. 5], if it is satisfied by all finite trees and is such that whenever a tree has this property, so do all the subtrees rooted at the children of the root. It is easily checked that any inherited property has probability either 0 or 1 given non-extinction. In particular, $\mathbf{P}\{D_\infty = 0 \,|\, \text{non-extinction}\}$ is either 0 or 1.

Theorem 5.2 (ii) says that $\mathbf{P}\{D_\infty = 0 \,|\, \text{non-extinction}\} = 0$ under the additional assumption (5.3).   □

The rest of the section is devoted to the proof of Theorem 5.2. Let us fix $\alpha \geq 0$. The idea is to use the spinal decomposition theorem in Example 4.6 of Sect. 4.7, associated with the positive harmonic function h on $[-\alpha, \infty)$. So it may be convenient to recall some basic ingredients in Example 4.6: (S_n) is the associated one-dimensional random walk, thus

$$\mathbf{E}[F(S_1 - S_0)] = \mathbf{E}\left[\sum_{|x|=1} F(V(x)) e^{-V(x)} \right],$$

for any measurable function $F : \mathbb{R} \to \mathbb{R}_+$, whereas R is the renewal function associated with (S_n) as in (A.2) of Appendix A.1, and the positive harmonic function h is defined as

$$h(u) := R(u + \alpha), \quad u \in [-\alpha, \infty).$$

Let $b \in [-\alpha, \infty)$. The new probability $\mathbf{Q}_b^{(\alpha)}$ (denoted by $\mathbf{Q}_b^{(h)}$ in Example 4.6 of Sect. 4.7) on $\mathscr{F}_\infty$ is such that

$$\mathbf{Q}_b^{(\alpha)}(A) = \int_A \frac{D_n^{(\alpha)}}{h(b)e^{-b}} \, d\mathbf{P}_b, \quad \forall A \in \mathscr{F}_n, \ \forall n \geq 0,$$

where

$$D_n^{(\alpha)} := \sum_{|x|=n} h(V(x))e^{-V(x)} \mathbf{1}_{\{V(y) \geq -\alpha, \ \forall y \in [\varnothing, x]\}}, \quad n \geq 0,$$

is a martingale under $\mathbf{P}_b$.

Recall that $\lim_{n \to \infty} \frac{R(u)}{u} = c_{\text{ren}} \in (0, \infty)$ (see (A.4) of Appendix A.1), so there exist constants $c_2 > 0$ and $c_3 > 0$ such that

$$c_2 (1 + u + \alpha) \leq h(u) \leq c_3 (1 + u + \alpha), \quad \forall u \geq -\alpha. \tag{5.7}$$

This inequality will be in use several times in the proof of Theorem 5.2.

Since $(D_n^{(\alpha)}, n \geq 0)$ is a non-negative martingale under $\mathbf{P} = \mathbf{P}_0$, it admits a finite $\mathbf{P}$-almost sure limit, denoted by $D_\infty^{(\alpha)}$, when $n \to \infty$. Let us first prove the following result.

Lemma 5.5 *Let $\alpha \geq 0$. Under Assumption (H), $D_n^{(\alpha)} \to D_\infty^{(\alpha)}$ in $L^1(\mathbf{P})$.*

Proof of Lemma 5.5 The main idea is already used in the proof of the Biggins martingale convergence theorem in Sect. 4.8, though the situation is a little more complicated here.

Let $\mathscr{G}_\infty^{(\alpha)}$ be the σ-field generated by $w_n^{(\alpha)}$ and $V(w_n^{(\alpha)})$ as well as the offspring of $w_n^{(\alpha)}$, for all $n \geq 0$. By Lemma 4.2 of Sect. 4.3, it suffices to show that

$$\liminf_{n \to \infty} \mathbf{E}_{\mathbf{Q}^{(\alpha)}}[D_n^{(\alpha)} \mid \mathscr{G}_\infty^{(\alpha)}] < \infty, \quad \mathbf{Q}^{(\alpha)}\text{-a.s.}$$

Using the martingale property of $D_n^{(\alpha)}$ for the subtrees rooted at the brothers of the spine, we arrive at (recalling that $\texttt{brot}(w_j^{(\alpha)})$ is the set of brothers of $w_j^{(\alpha)}$):

$$\mathbf{E}_{\mathbf{Q}^{(\alpha)}}[D_n^{(\alpha)} \mid \mathscr{G}_\infty^{(\alpha)}] = h(V(w_n^{(\alpha)}))e^{-V(w_n^{(\alpha)})}$$

$$+ \sum_{k=1}^n \sum_{x \in \texttt{brot}(w_j^{(\alpha)})} h(V(x))e^{-V(x)} \mathbf{1}_{\{V(x_j) \geq -\alpha, \ \forall j \leq k\}}.$$

We let $n \to \infty$. The term $h(V(w_n^{(\alpha)}))e^{-V(w_n^{(\alpha)})}$ is easily treated: Since $(V(w_n^{(\alpha)}), n \geq 0)$ under $\mathbf{Q}^{(\alpha)}$ is a centred random walk conditioned to stay non-negative, we have $V(w_n^{(\alpha)}) \to \infty$, $\mathbf{Q}^{(\alpha)}$-a.s., therefore $h(V(w_n^{(\alpha)}))e^{-V(w_n^{(\alpha)})} \to 0$, $\mathbf{Q}^{(\alpha)}$-a.s. On the other hand, we simply use (5.7) to say that $h(V(x)) \mathbf{1}_{\{V(x_j) \geq -\alpha, \, \forall j \leq k\}}$ is bounded by $c_3 [1 + (\alpha + V(x))_+]$. Therefore,

$$\liminf_{n \to \infty} \mathbf{E}_{\mathbf{Q}^{(\alpha)}}[D_n^{(\alpha)} \mid \mathscr{G}_\infty^{(\alpha)}] \leq c_3 \sum_{k=1}^{\infty} \sum_{x \in \mathrm{brot}(w_j^{(\alpha)})} [1 + (\alpha + V(x))_+]e^{-V(x)}.$$

It remains to show that the right-hand side is $\mathbf{Q}^{(\alpha)}$-a.s. finite. Using the trivial inequality $1 + (\alpha + V(x))_+ \leq [1 + \alpha + V(w_{k-1}^{(\alpha)})] + [V(x) - V(w_{k-1}^{(\alpha)})]_+$, we only need to check that $\mathbf{Q}^{(\alpha)}$-almost surely,

$$\sum_{k=1}^{\infty}[1 + \alpha + V(w_{k-1}^{(\alpha)})]e^{-V(w_{k-1}^{(\alpha)})} \sum_{x \in \mathrm{brot}(w_j^{(\alpha)})} e^{-[V(x) - V(w_{k-1}^{(\alpha)})]} < \infty, \qquad (5.8)$$

$$\sum_{k=1}^{\infty} e^{-V(w_{k-1}^{(\alpha)})} \sum_{x \in \mathrm{brot}(w_j^{(\alpha)})} [V(x) - V(w_{k-1}^{(\alpha)})]_+ \, e^{-[V(x) - V(w_{k-1}^{(\alpha)})]} < \infty. \qquad (5.9)$$

Recall the definitions: $X = \sum_{|x|=1} e^{-V(x)}$ and $\widetilde{X} = \sum_{|x|=1} V(x)_+ e^{-V(x)}$ (see (5.4)). Recall from (5.7) that $h(u) \mathbf{1}_{\{u \geq -\alpha\}} \leq c_3 [1 + (u + \alpha)_+]$. Let $b \geq -\alpha$. Since $1 + (u + \alpha)_+ \leq [1 + b + \alpha] + (u - b)_+$, we have

$$D_1^{(\alpha)} = \sum_{|x|=1} h(V(x))e^{-V(x)} \mathbf{1}_{\{V(x) \geq -\alpha\}}$$

$$\leq c_3 \sum_{|x|=1} e^{-V(x)} \{[1 + b + \alpha] + [V(x) - b]_+\}.$$

Therefore, for any $z \in \mathbb{R}$ and $b \geq -\alpha$,

$$\mathbf{Q}_b^{(\alpha)} \left\{ \sum_{|x|=1} e^{-[V(x) - b]} > z \right\}$$

$$= \frac{1}{h(b)e^{-b}} \mathbf{E}_b \left[D_1^{(\alpha)} \mathbf{1}_{\{\sum_{|x|=1} e^{-[V(x)-b]} > z\}} \right]$$

$$\leq \frac{c_3}{c_2} e^b \mathbf{E}_b \left[\sum_{|y|=1} e^{-V(y)} (1 + \frac{(V(y) - b)_+}{1 + b + \alpha}) \mathbf{1}_{\{\sum_{|x|=1} e^{-[V(x)-b]} > z\}} \right]$$

$$= c_4 \, \mathbf{E}[X \mathbf{1}_{\{X > z\}}] + \frac{c_4}{1 + b + \alpha} \mathbf{E}[\widetilde{X} \mathbf{1}_{\{X > z\}}], \qquad (5.10)$$

with $c_4 := \frac{c_3}{c_2}$. For later use, we note that the same argument gives that for $z \in \mathbb{R}$ and $b \geq -\alpha$,

$$\mathbf{Q}_b^{(\alpha)} \left\{ \sum_{|x|=1} (V(x) - b)_+ \, e^{-[V(x)-b]} > z \right\}$$

$$\leq c_4 \, \mathbf{E}[X \, \mathbf{1}_{\{\widetilde{X}>z\}}] + \frac{c_4}{1+b+\alpha} \, \mathbf{E}[\widetilde{X} \, \mathbf{1}_{\{\widetilde{X}>z\}}]. \qquad (5.11)$$

We take $z := e^{\lambda b}$ in (5.10), where $\lambda \in (0, 1)$ is a fixed real number. Writing $f_1(z) := \mathbf{E}[X \, \mathbf{1}_{\{X>z\}}]$ and $f_2(z) := \mathbf{E}[\widetilde{X} \, \mathbf{1}_{\{X>z\}}]$ for all $z \in \mathbb{R}$, it follows from the Markov property (applied at time $k - 1$) that

$$\mathbf{Q}^{(\alpha)} \left\{ \sum_{x \in \mathrm{brot}(w_k^{(\alpha)})} e^{-[V(x)-V(w_{k-1}^{(\alpha)})]} > e^{\lambda V(w_{k-1}^{(\alpha)})} \right\}$$

$$\leq c_4 \, \mathbf{E}_{\mathbf{Q}^{(\alpha)}} \left[f_1(e^{\lambda V(w_{k-1}^{(\alpha)})}) + \frac{f_2(e^{\lambda V(w_{k-1}^{(\alpha)})})}{1 + V(w_{k-1}^{(\alpha)}) + \alpha} \right].$$

We now estimate the expectation term on the right-hand side. Recall from the spinal decomposition theorem that under $\mathbf{Q}^{(\alpha)}$, $(V(w_i^{(\alpha)}), \, i \geq 0)$ is a random walk conditioned to stay in $[-\alpha, \infty)$ in the sense of Doob's h-transform (see (4.4) of Sect. 4.2), so

$$\mathbf{E}_{\mathbf{Q}^{(\alpha)}} \left[f_1(e^{\lambda V(w_{k-1}^{(\alpha)})}) + \frac{f_2(e^{\lambda V(w_{k-1}^{(\alpha)})})}{1 + V(w_{k-1}^{(\alpha)}) + \alpha} \right]$$

$$= \frac{1}{h(0)} \, \mathbf{E} \left[\left(f_1(e^{\lambda S_{k-1}}) + \frac{f_2(e^{\lambda S_{k-1}})}{1 + S_{k-1} + \alpha} \right) h(S_{k-1}) \, \mathbf{1}_{\{\min_{0 \leq i \leq k-1} S_i \geq -\alpha\}} \right]$$

$$= \frac{1}{h(0)} \, \mathbf{E} \left[\left(X \mathbf{1}_{\{S_{k-1} \leq \frac{1}{\lambda} \ln X\}} + \frac{\widetilde{X} \, \mathbf{1}_{\{S_{k-1} \leq \frac{1}{\lambda} \ln X\}}}{1 + S_{k-1} + \alpha} \right) h(S_{k-1}) \, \mathbf{1}_{\{\min_{0 \leq i \leq k-1} S_i \geq -\alpha\}} \right],$$

where the random walk $(S_i, \, i \geq 0)$ is taken to be independent of the pair $(X, \widetilde{X})$. Applying again (5.7) yields that

$$\mathbf{E}_{\mathbf{Q}^{(\alpha)}} \left[f_1(e^{\lambda V(w_{k-1}^{(\alpha)})}) + \frac{f_2(e^{\lambda V(w_{k-1}^{(\alpha)})})}{1 + V(w_{k-1}^{(\alpha)}) + \alpha} \right]$$

$$\leq \frac{c_3}{h(0)} \, \mathbf{E} \left[\left((1 + \frac{\ln X}{\lambda} + \alpha) X \mathbf{1}_{\{S_{k-1} \leq \frac{1}{\lambda} \ln X\}} + \widetilde{X} \mathbf{1}_{\{S_{k-1} \leq \frac{1}{\lambda} \ln X\}} \right) \mathbf{1}_{\{\min_{0 \leq i \leq k-1} S_i \geq -\alpha\}} \right].$$

Recall from Lemma A.5 (Appendix A.2) that $\sum_{\ell=0}^{\infty} \mathbf{P}\{S_\ell \leq y - z, \, \min_{0 \leq i \leq \ell} S_i \geq -z\} \leq c_5 \, (1 + y)(1 + \min\{y, z\})$ for some constant $c_5 > 0$ and all $y \geq 0$ and $z \geq 0$.

This implies

$$\sum_{k=1}^{\infty} \mathbf{E}_{\mathbf{Q}^{(\alpha)}} \left[f_1(e^{\lambda V(w_{k-1}^{(\alpha)})}) + \frac{f_2(e^{\lambda V(w_{k-1}^{(\alpha)})})}{1 + V(w_{k-1}^{(\alpha)}) + \alpha} \right]$$

$$\leq \frac{c_6}{h(0)} \mathbf{E}[X (1 + \ln_+ X)^2 + \widetilde{X} (1 + \ln_+ X)],$$

which is finite under assumption (5.3) (using also the second part of its consequence (5.5)). We have thus proved that

$$\sum_{k=1}^{\infty} \mathbf{Q}^{(\alpha)} \left\{ \sum_{x \in \text{brot}(w_k^{(\alpha)})} e^{-[V(x) - V(w_{k-1}^{(\alpha)})]} > e^{\lambda V(w_{k-1}^{(\alpha)})} \right\} < \infty.$$

By the Borel–Cantelli lemma, for any $\lambda \in (0, 1)$, $\mathbf{Q}^{(\alpha)}$-almost surely for all sufficiently large k,

$$\sum_{x \in \text{brot}(w_k^{(\alpha)})} e^{-[V(x) - V(w_{k-1}^{(\alpha)})]} \leq e^{\lambda V(w_{k-1}^{(\alpha)})}.$$

Since $V(w_{k-1}^{(\alpha)}) \geq (k-1)^{(1/2)-\varepsilon}$ for any $\varepsilon > 0$ and $\mathbf{Q}^{(\alpha)}$-almost surely all sufficiently large k (this is a known property of conditioned random walks; see Biggins [53]), we obtain (5.8).

The proof of (5.9) is along the same lines, except that we use (5.11) instead of (5.10), and $\mathbf{E}[X (1 + \ln_+ \widetilde{X})^2 + \widetilde{X} (1 + \ln_+ \widetilde{X})] < \infty$ instead of $\mathbf{E}[X (1 + \ln_+ X)^2 + \widetilde{X} (1 + \ln_+ X)] < \infty$. This completes the proof of Lemma 5.5. □

We have now all the necessary ingredients to prove Theorem 5.2.

Proof of Theorem 5.2 (i) Assume $\psi(0) > 0$, $\psi(1) = 0 = \psi'(1)$. Assume also $\mathbf{E}[\sum_{|x|=1} V(x)_-^2 e^{-V(x)}] < \infty$. We only need to prove the almost sure convergence of D_n (see Discussion 5.3).

Fix $\varepsilon > 0$. Recall that $\mathbf{P}^*(\cdot) := \mathbf{P}(\cdot \mid \text{non-extinction})$. Since $\psi(1) = 0$, it follows from Lemma 3.1 of Sect. 3.1 that

$$\inf_{|x|=n} V(x) \to \infty, \quad \inf_{x \in \mathbb{T}} V(x) > -\infty, \quad \mathbf{P}^*\text{-a.s.} \tag{5.12}$$

We can thus fix $\alpha = \alpha(\varepsilon) \geq 0$ such that $\mathbf{P}\{\inf_{x \in \mathbb{T}} V(x) > -\alpha\} \geq 1 - \varepsilon$.

Consider $D_n^{(\alpha)} := \sum_{|x|=n} h(V(x)) e^{-V(x)} \mathbf{1}_{\{V(y) \geq -\alpha, \forall y \in [\emptyset, x]\}}$, which is a non-negative martingale. On the one hand, it converges a.s. to $D_\infty^{(\alpha)}$; on the other hand, on the set $\{\inf_{x \in \mathbb{T}} V(x) > -\alpha\}$, $D_n^{(\alpha)}$ coincides with $\sum_{|x|=n} h(V(x)) e^{-V(x)}$, which, $\mathbf{P}^*$-a.s., is equivalent to $c_{\text{ren}} (D_n + \alpha W_n)$ (when $n \to \infty$) under the assumption $\mathbf{E}[\sum_{|x|=1} V(x)_-^2 e^{-V(x)}] < \infty$ (see (A.5) in Appendix A.1). Since $W_n \to 0$ a.s., we obtain that with probability at least $1 - \varepsilon$, D_n converges to a finite limit. This yields the almost sure convergence of D_n.

(ii) Let us now work under Assumption (H). According to Discussion 5.4, we only need to check that $\mathbf{P}\{D_\infty > 0\} > 0$.

Let $\alpha = 0$ and consider $D_n^{(0)} := \sum_{|x|=n} h(V(x)) e^{-V(x)} \mathbf{1}_{\{V(y) \geq 0, \, \forall y \in [\varnothing, x]\}}$. By Lemma 5.5, $D_n^{(0)} \to D_\infty^{(0)}$ in L^1, so $\mathbf{E}(D_\infty^{(0)}) = 1$. In particular, $\mathbf{P}\{D_\infty^{(0)} > 0\} > 0$.

On the other hand, $h(u) \leq c_3 (1 + u)$ for $u \geq 0$ (see (5.7); recalling that $\alpha = 0$), so $D_n^{(0)} \leq c_3 (W_n + \sum_{|x|=n} V(x)_+ e^{-V(x)})$, which converges a.s. to $c_3 (0 + D_\infty) = c_3 D_\infty$, we obtain $D_\infty^{(0)} \leq c_3 D_\infty$; hence $\mathbf{P}\{D_\infty > 0\} > 0$. □

One may wonder whether or not the assumptions in Theorem 5.2 are optimal, respectively, for the a.s. convergence of D_n, and for the positivity of the limit. No definitive answer is, alas, available so far. Concerning the a.s. convergence of D_n, we ask:

Question 5.6 Assume $\psi(0) > 0$ and $\psi(1) = 0 = \psi'(1)$. Does D_n converge almost surely?

The answer to Question 5.6 should be negative. A correct formulation of the question is: how to weaken the conditions in Theorem 5.2 to ensure almost sure convergence of D_n?

For the a.s. positivity of D_∞, it looks natural to wonder whether the assumption $\mathbf{E}[\sum_{|x|=1} V(x)^2 e^{-V(x)}] < \infty$ could be weakened to $\mathbf{E}[\sum_{|x|=1} V(x)_-^2 e^{-V(x)}] < \infty$ in Lemma 5.5, which leads to the following question:

Question 5.7 Does Theorem 5.2 *(ii)* hold if we assume $\mathbf{E}[\sum_{|x|=1} V(x)_-^2 e^{-V(x)}] < \infty$ instead of $\mathbf{E}[\sum_{|x|=1} V(x)^2 e^{-V(x)}] < \infty$?

I feel that the answer to Question 5.7 should be negative as well.

If $\mathbf{E}[\sum_{|x|=1} V(x)^2 e^{-V(x)}] < \infty$, assumption (5.3) is likely to be also necessary to ensure the positivity of D_∞. Let us recall the following

Theorem 5.8 (Biggins and Kyprianou [57]) *Assume* $\psi(0) > 0$, $\psi(1) = 0 = \psi'(1)$ *and* $\mathbf{E}[\sum_{|x|=1} V(x)^2 e^{-V(x)}] < \infty$. *If either* $\mathbf{E}[\frac{X \ln_+^2 X}{\ln_+ \ln_+ \ln_+ X}] = \infty$ *or* $\mathbf{E}[\frac{\widetilde{X} \ln_+ \widetilde{X}}{\ln_+ \ln_+ \ln_+ \widetilde{X}}] = \infty$, *then* $D_\infty = 0$, $\mathbf{P}^*$-*a.s.*

The presence of $\ln_+ \ln_+ \ln_+$ terms in Theorem 5.8, originating in [57] from the oscillations of the branching random walk along the spine (which is a one-dimensional centred random walk conditioned to stay non-negative), is expected to be superfluous. This leads to the following

Conjecture 5.9 Assume $\psi(0) > 0$, $\psi(1) = 0 = \psi'(1)$, $\mathbf{E}[\sum_{|x|=1} V(x)^2 e^{-V(x)}] < \infty$. Then (5.3) is a necessary and sufficient condition for $\mathbf{P}\{D_\infty > 0\} > 0$.

According to Discussion 5.3, $\mathbf{P}^*\{D_\infty > 0\}$ is either 0 or 1, so $\mathbf{P}\{D_\infty > 0\} > 0$ means $D_\infty > 0$, $\mathbf{P}^*$-a.s.

We mention that a necessary and sufficient condition for the positivity of D_∞ is obtained by Biggins and Kyprianou [58] in terms of the asymptotic behaviour near the origin of the Laplace transform of the fixed point of the associated smoothing transform (defined in (3.1) of Sect. 3.3). The analogue of Conjecture 5.9 for branching Brownian motion is proved by Ren and Yang [213].

[N.B.: Since the preparation of the first draft of these notes, Conjecture 5.9 has been solved, in the affirmative, by Chen [88].]

We close this section with a couple of remarks.

Remark 5.10 Under Assumption (H), the derivative martingale has a $\mathbf{P}^*$-a.s. positive limit, whereas the additive martingale tends $\mathbf{P}^*$-a.s. to 0, so the advantage of using the derivative martingale[3] while applying the spinal decomposition theorem is clear: It allows us to have a probability measure which is absolutely continuous with respect to $\mathbf{P}$.   □

Remark 5.11 Assume $\psi(0) > 0$ and $\psi(1) = 0 = \psi'(1)$. Theorem 5.2(i) tells us that if $\mathbf{E}[\sum_{|x|=1} V(x)_-^2 \, e^{-V(x)}] < \infty$, then $D_n \to D_\infty$ a.s. The derivative martingale D_n sums over all the particles in the n-th generation. Sometimes, however, it is useful to know whether the convergence still holds if we sum over particles belonging to some special random collections. This problem is studied in Biggins and Kyprianou [57]. Their result applies when these special random collections are the so-called stopping lines. Let us record here a particularly useful consequence: Let[4]

$$\mathscr{Z}[A] := \left\{ x \in \mathbb{T} : V(x) \geq A, \ \max_{0 \leq i < |x|} V(x_i) < A \right\} \tag{5.13}$$

then under the assumptions of Theorem 5.2(i), i.e., if $\psi(0) > 0$, $\psi(1) = 0 = \psi'(1)$ and $\mathbf{E}[\sum_{|x|=1} V(x)_-^2 \, e^{-V(x)}] < \infty$, then as $A \to \infty$,

$$\sum_{x \in \mathscr{Z}[A]} V(x) e^{-V(x)} \to D_\infty, \quad \text{a.s.}$$

Moreover, $\sum_{x \in \mathscr{Z}[A]} e^{-V(x)} \to 0$ a.s.   □

5.3 Leftmost Position: Weak Convergence

We work under Assumption (H) (see Sect. 5.1).

One of the main concerns of this chapter is to prove Theorem 5.15 in Sect. 5.4, saying that if the law of the underlying point process $\varXi$ is non-lattice, then on the

---

[3] Strictly speaking, we use its approximation $D_n^{(\alpha)}$ in order to get a positive measure.

[4] In Biggins and Kyprianou [57], the set $\mathscr{Z}[A]$ is called a "very simple optional line".

set of non-extinction, $\inf_{|x|=n} V(x) - \frac{3}{2} \ln n$ converges weakly to a non-degenerate limiting distribution. As a warm up to the highly technical proof of this deep result, we devote this section to explaining why $\frac{3}{2} \ln n$ should be the correct centering term. Recall that $\mathbf{P}^*(\cdot) := \mathbf{P}(\cdot \mid \text{non-extinction})$.

Theorem 5.12 *Under Assumption* (H),

$$\frac{1}{\ln n} \inf_{|x|=n} V(x) \to \frac{3}{2}, \quad \text{in probability, under } \mathbf{P}^*.$$

The proof of the theorem relies on the following preliminary lemma, which is stated uniformly in $z \in [0, \frac{3}{2} \ln n]$ for an application in Sect. 5.4.2. For the proof of Theorem 5.12, only the case $z = 0$ is needed.

Let $C > 0$ be the constant in Lemma A.10 (Appendix A.2).

Lemma 5.13 *Under Assumption* (H),

$$\liminf_{n\to\infty} \inf_{z\in[0, \frac{3}{2} \ln n]} e^z \mathbf{P}\Big\{\exists x \in \mathbb{T} : |x| = n,$$

$$\min_{1\le i\le n} V(x_i) \ge 0, \ \frac{3}{2} \ln n - z \le V(x) \le \frac{3}{2} \ln n - z + C\Big\} > 0.$$

We first admit Lemma 5.13, and proceed to the proof of the theorem.

Proof of Theorem 5.12 The proof is carried out in two steps.

First step. For any $\varepsilon > 0$,

$$\mathbf{P}^*\Big\{\inf_{|x|=n} V(x) \ge (\frac{3}{2} - \varepsilon) \ln n\Big\} \to 1, \quad n \to \infty. \tag{5.14}$$

To prove (5.14), we fix a constant $\alpha > 0$, and use the many-to-one formula (Theorem 1.1 in Sect. 1.3) to see that

$$\mathbf{E}\Big(\sum_{|x|=n} \mathbf{1}_{\{V(x)\le(\frac{3}{2}-\varepsilon)\ln n, \ V(x_i)\ge-\alpha, \ \forall 1\le i\le n\}}\Big)$$

$$= \mathbf{E}\Big[e^{S_n} \mathbf{1}_{\{S_i\le(\frac{3}{2}-\varepsilon)\ln n, \ S_i\ge-\alpha, \ \forall 1\le i\le n\}}\Big]$$

$$\le n^{\frac{3}{2}-\varepsilon} \mathbf{P}\Big(S_i \le (\frac{3}{2} - \varepsilon) \ln n, \ S_i \ge -\alpha, \ \forall 1 \le i \le n\Big).$$

As long as n is sufficiently large such that $(\frac{3}{2} - \varepsilon) \ln n \ge 1$, we have $\mathbf{P}\{S_i \le (\frac{3}{2} - \varepsilon) \ln n, \ S_i \ge -\alpha, \ \forall 1 \le i \le n\} \le c_7 \frac{(\ln n)^2}{n^{3/2}}$ with $c_7 = c_7(\alpha, \varepsilon)$ (by Lemma A.1 of Appendix A.2). It follows that

$$\lim_{n\to\infty} \mathbf{E}\Big(\sum_{|x|=n} \mathbf{1}_{\{V(x)\le(\frac{3}{2}-\varepsilon)\ln n, \ V(x_i)\ge-\alpha, \ \forall 1\le i\le n\}}\Big) = 0.$$

Since $\inf_{x\in\mathbb{T}} V(x) > -\infty$ a.s. (see (5.12)), (5.14) follows. It is worth noting that (5.14) holds without assumption (5.3).

Second step. For any $\varepsilon > 0$,

$$\mathbf{P}^*\Big\{\inf_{|x|=n} V(x) \le (\frac{3}{2}+\varepsilon)\ln n\Big\} \to 1, \quad n\to\infty. \tag{5.15}$$

By Lemma 5.13, there exists $\delta > 0$ such that for all sufficiently large n, say $n \ge n_0$,

$$\mathbf{P}\Big\{\inf_{|x|=n} V(x) \le \frac{3}{2}\ln n + C\Big\} \ge \delta.$$

Let $\varepsilon > 0$. Let $\ell \ge 1$ be an integer such that $(1-\delta)^\ell \le \varepsilon$. The system surviving $\mathbf{P}^*$-a.s., there exist $A > 0$ and an integer $n_1 \ge 1$ such that $\mathbf{P}^*(\sum_{|x|=n_1} \mathbf{1}_{\{V(x)\le A\}} \ge \ell) \ge 1-\varepsilon$. By considering all the subtrees rooted at the vertices x of generation n_1 with $V(x) \le A$, we see that for all $n \ge n_0$,

$$\mathbf{P}^*\Big\{\inf_{|x|=n+n_1} V(x) \le \frac{3}{2}\ln n + C + A\Big\} \ge 1-(1-\delta)^\ell - \varepsilon \ge 1-2\varepsilon,$$

which yields (5.15). □

We now turn to the proof of Lemma 5.13, by applying the spinal decomposition theorem to the size-biased branching random walk as seen in Example 4.5 of Sect. 4.6: Let $W_n := \sum_{|x|=n} \mathrm{e}^{-V(x)}$ be the additive martingale, and let $\mathbf{Q}$ be such that

$$\mathbf{Q}(A) = \int_A W_n \,\mathrm{d}\mathbf{P}, \quad \forall A \in \mathscr{F}_n, \ \forall n \ge 0.$$

Under $\mathbf{Q}$, $(V(w_n))$ is a centred random walk.

Let $L \ge 0$ and $0 \le K \le z \le \frac{3}{2}\ln n - L$. Let

$$a_i^{(n)} = a_i^{(n)}(z, L, K) := \begin{cases} -z + K & \text{if } 0 \le i \le \lfloor \frac{n}{2} \rfloor, \\ \frac{3}{2}\ln n - z - L & \text{if } \lfloor \frac{n}{2} \rfloor < i \le n. \end{cases} \tag{5.16}$$

We consider[5]

$$\mathscr{Z}_n^{z,L,K} := \Big\{x:\ |x| = n,\ V(x_i) \ge a_i^{(n)},\ \forall 0 \le i \le n,$$
$$V(x) \le \frac{3}{2}\ln n - z + C\Big\}. \tag{5.17}$$

[5]We prepare for a general result, see the so-called peeling lemma for the spine, stated as Theorem 5.14 below, which will be applied to various situations. For the proof of Lemma 5.13, we simply take $L = 0$ and later $K = z$; for the proof of Theorem 5.12, we moreover take $z = 0$.

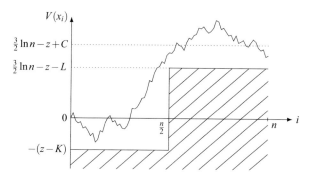

Fig. 5.1 Absorbing barrier in $\mathscr{L}_n^{z,L,K}$

In $\mathscr{L}_n^{z,L,K}$, we **add an absorbing barrier**, killing all vertices until generation n whose spatial value is below $-z+K$ as well as all vertices between generation $\frac{n}{2}$ and generation n whose spatial value is below $\frac{3}{2}\ln n - z - L$. See Fig. 5.1 for an example of vertex $x \in \mathscr{L}_n^{z,L,K}$.

Let $\#\mathscr{L}_n^{z,L,K}$ denote the cardinality of $\mathscr{L}_n^{z,L,K}$. By definition,

$$\mathbf{E}(\#\mathscr{L}_n^{z,L,K}) = \mathbf{E}_{\mathbf{Q}}\left[\frac{\#\mathscr{L}_n^{z,L,K}}{W_n}\right] = \mathbf{E}_{\mathbf{Q}}\left[\sum_{|x|=n}\frac{\mathbf{1}_{\{x\in\mathscr{L}_n^{z,L,K}\}}}{W_n}\right].$$

Since $\mathbf{Q}\{w_n = x \mid \mathscr{F}_n\} = \frac{e^{-V(x)}}{W_n}$ for any $x \in \mathbb{T}$ with $|x| = n$, we have

$$\mathbf{E}(\#\mathscr{L}_n^{z,L,K}) = \mathbf{E}_{\mathbf{Q}}\left[\sum_{|x|=n}\mathbf{1}_{\{x\in\mathscr{L}_n^{z,L,K}\}}\,e^{V(x)}\,\mathbf{1}_{\{w_n=x\}}\right] = \mathbf{E}_{\mathbf{Q}}\left[e^{V(w_n)}\,\mathbf{1}_{\{w_n\in\mathscr{L}_n^{z,L,K}\}}\right].$$

By definition, $\frac{3}{2}\ln n - z - L \le V(w_n) \le \frac{3}{2}\ln n - z + C$ on $\{w_n \in \mathscr{L}_n^{z,L,K}\}$, so

$$n^{3/2}\,e^{-z-L}\,\mathbf{Q}(w_n \in \mathscr{L}_n^{z,L,K}) \le \mathbf{E}(\#\mathscr{L}_n^{z,L,K}) \le n^{3/2}\,e^{-z+C}\,\mathbf{Q}(w_n \in \mathscr{L}_n^{z,L,K}).$$

The spinal decomposition theorem stating that the process $(V(w_n))_{n\ge0}$ under $\mathbf{Q}$ has the law of $(S_n)_{n\ge0}$ under $\mathbf{P}$, we obtain, by Lemma A.10 (Appendix A.2) for the lower bound and by Lemma A.4 (Appendix A.2) for the upper bound, that for $n \ge n_0$,

$$\mathbf{Q}\left(w_n \in \mathscr{L}_n^{z,L,K}\right) = \mathbf{P}\left\{S_i \ge a_i^{(n)},\ \forall 0 \le i \le n,\ S_n \le \frac{3}{2}\ln n - z + C\right\}$$

$$\in \left[c_8\,\frac{1+z-K}{n^{3/2}},\ c_9\,\frac{1+z-K}{n^{3/2}}\right], \tag{5.18}$$

with $c_8 = c_8(L) > 0$ and $c_9 = c_9(L) > 0$. In particular,

$$\mathbf{E}(\#\mathscr{L}_n^{z,L,K}) \geq c_8 \, (1 + z - K) \, e^{-z-L}.$$

Unfortunately, the second moment of $\#\mathscr{L}_n^{z,L,K}$ turns out to be large. As before, the idea is to restrict ourselves to a suitably chosen subset of $\mathscr{L}_n^{z,L,K}$. Let, for $n \geq 2$, $L \geq 0$ and $0 \leq K \leq z \leq \frac{3}{2} \ln n - L$,

$$\beta_i^{(n)} := \begin{cases} i^{1/7} & \text{if } 0 \leq i \leq \lfloor \frac{n}{2} \rfloor, \\ (n-i)^{1/7} & \text{if } \lfloor \frac{n}{2} \rfloor < i \leq n. \end{cases} \tag{5.19}$$

[The power $1/7$ is chosen arbitrarily; anything lying in $(0, \frac{1}{6})$ will do the job; it originates from Lemma A.6 of Appendix A.2.] Consider

$$\mathscr{Y}_n^{z,L,K} := \Bigl\{ |x| = n : \sum_{y \in \text{brot}(x_{i+1})} [1 + (V(y) - a_i^{(n)})_+] e^{-(V(y) - a_i^{(n)})} \leq \rho \, e^{-\beta_i^{(n)}}, $$

$$\forall 0 \leq i \leq n-1 \Bigr\}, \tag{5.20}$$

with $\text{brot}(x_{i+1})$ denoting, as before, the set of brothers of x_{i+1}, which is possibly empty.

The following peeling lemma for the spine allows to throw away negligible events and to control the second moment in various settings. The constant $C > 0$, in the definition of $\mathscr{L}_n^{z,L,K}$ (see (5.17)), was chosen at the beginning of the section to be the one in Lemma A.10 (Appendix A.2) in order to guarantee the *lower* bound in (5.18). The peeling lemma for the spine, stated below, provides an *upper* bound for $\mathbf{Q}(w_n \in \mathscr{L}_n^{z,L,K} \setminus \mathscr{Y}_n^{z,L,K})$, and does not have any special requirement for $C > 0$.

Theorem 5.14 (The Peeling Lemma) *Let $L \geq 0$ and $C > 0$. For any $\varepsilon > 0$, we can choose the constant $\rho > 0$ in (5.20) to be sufficiently large, such that for all sufficiently large n, and all $0 \leq K \leq z \leq \frac{3}{2} \ln n - L$,*

$$\mathbf{Q}\Bigl(w_n \in \mathscr{L}_n^{z,L,K} \setminus \mathscr{Y}_n^{z,L,K}\Bigr) \leq \varepsilon \, \frac{1 + z - K}{n^{3/2}},$$

where $\mathscr{L}_n^{z,L,K}$ and $\mathscr{Y}_n^{z,L,K}$ are defined in (5.17) and (5.20), respectively.

The proof of the peeling lemma is postponed until Sect. 5.8.
We are now able to prove Lemma 5.13.

Proof of Lemma 5.13 Let $C > 0$ be the constant in Lemma A.10 of Appendix A.2. Let n_0 satisfy (5.18). Let $n \geq n_0$ and let $0 \leq K \leq z \leq \frac{3}{2} \ln n$. Let $\mathscr{L}_n^{z,0,K}$ be as in (5.17) with $L = 0$, and $\mathscr{Y}_n^{z,0,K}$ as in (5.20) with $L = 0$.

By (5.18), $c_8 \frac{1+z-K}{n^{3/2}} \leq \mathbf{Q}(w_n \in \mathscr{L}_n^{z,0,K}) \leq c_9 \frac{1+z-K}{n^{3/2}}$. On the other hand, by the peeling lemma (Theorem 5.14), it is possible to choose the constant $\rho > 0$ in the event $\mathscr{Y}_n^{z,0,K}$ defined in (5.20) such that $\mathbf{Q}(w_n \in \mathscr{L}_n^{z,0,K} \setminus \mathscr{Y}_n^{z,0,K}) \le \frac{c_8}{2} \frac{1+z-K}{n^{3/2}}$, uniformly in $z \in [0, \frac{3}{2} \ln n]$. As such,

$$\frac{c_8}{2} \frac{1+z-K}{n^{3/2}} \le \mathbf{Q}(w_n \in \mathscr{L}_n^{z,0,K} \cap \mathscr{Y}_n^{z,0,K}) \le c_9 \frac{1+z-K}{n^{3/2}}, \tag{5.21}$$

from which it follows that

$$\mathbf{E}[\zeta_n(z)] \ge c_{10} \, (1+z-K) \, e^{-z},$$

where

$$\zeta_n(z) := \#(\mathscr{L}_n^{z,0,K} \cap \mathscr{Y}_n^{z,0,K}).$$

[The constant c_{10} depends on the fixed parameter C.]

We now estimate the second moment of $\zeta_n(z)$. Using again the probability $\mathbf{Q}$, we have,

$$\mathbf{E}[\zeta_n(z)^2] = \mathbf{E}_{\mathbf{Q}}\Big[\#(\mathscr{L}_n^{z,0,K} \cap \mathscr{Y}_n^{z,0,K}) \sum_{|x|=n} \frac{\mathbf{1}_{\{x \in \mathscr{L}_n^{z,0,K} \cap \mathscr{Y}_n^{z,0,K}\}}}{W_n}\Big]$$

$$= \mathbf{E}_{\mathbf{Q}}\Big[\#(\mathscr{L}_n^{z,0,K} \cap \mathscr{Y}_n^{z,0,K}) \, e^{V(w_n)} \, \mathbf{1}_{\{w_n \in \mathscr{L}_n^{z,0,K} \cap \mathscr{Y}_n^{z,0,K}\}}\Big].$$

Accordingly,

$$\mathbf{E}[\zeta_n(z)^2] \le n^{3/2} \, e^{-z+C} \, \mathbf{E}_{\mathbf{Q}}\Big[\#(\mathscr{L}_n^{z,0,K} \cap \mathscr{Y}_n^{z,0,K}) \, \mathbf{1}_{\{w_n \in \mathscr{L}_n^{z,0,K} \cap \mathscr{Y}_n^{z,0,K}\}}\Big]$$

$$\le n^{3/2} \, e^{-z+C} \, \mathbf{E}_{\mathbf{Q}}\Big[(\#\mathscr{L}_n^{z,0,K}) \, \mathbf{1}_{\{w_n \in \mathscr{L}_n^{z,0,K} \cap \mathscr{Y}_n^{z,0,K}\}}\Big].$$

Decomposing $\#\mathscr{L}_n^{z,0,K}$ along the spine yields that

$$\#\mathscr{L}_n^{z,0,K} = \mathbf{1}_{\{w_n \in \mathscr{L}_n^{z,0,K}\}} + \sum_{i=1}^{n} \sum_{y \in \mathrm{brot}(w_i)} \#(\mathscr{L}_n^{z,0,K}(y)),$$

where $\mathrm{brot}(w_i)$ is, as before, the set of brothers of w_i, and $\mathscr{L}_n^{z,0,K}(y) := \{x \in \mathscr{L}_n^{z,0,K} : x \ge y\}$ the set of descendants x of y at generation n such that $x \in \mathscr{L}_n^{z,0,K}$. By the spinal decomposition theorem, conditioning on $\mathscr{G}_\infty$, the σ-field generated by w_j and $V(w_j)$ as well as the offspring of w_j, for all $j \ge 0$, we have, for $y \in \mathrm{brot}(w_i)$,

$$\mathbf{E}_{\mathbf{Q}}\Big[\#\mathscr{L}_n^{z,0,K}(y) \,|\, \mathscr{G}_\infty\Big] = \varphi_{i,n}(V(y)),$$

where, for $r \in \mathbb{R}$,

$$\varphi_{i,n}(r) := \mathbf{E}\left[\sum_{|x|=n-i} \mathbf{1}_{\{r+V(x_j)\geq a_{j+i}^{(n)}, \forall 0 \leq j \leq n-i, \, r+V(x) \leq \frac{3}{2}\ln n - z + C\}}\right]$$

$$= \mathbf{E}\left[e^{S_{n-i}} \mathbf{1}_{\{r+S_j \geq a_{j+i}^{(n)}, \forall 0 \leq j \leq n-i, \, r+S_{n-i} \leq \frac{3}{2}\ln n - z + C\}}\right]$$

$$\leq n^{3/2} e^{C-z-r} \mathbf{P}\left(r + S_j \geq a_{j+i}^{(n)}, \, \forall 0 \leq j \leq n-i,\right.$$

$$\left. r + S_{n-i} \leq \frac{3}{2}\ln n - z + C\right), \tag{5.22}$$

and (S_j) is the associated random walk, the last identity being a consequence of the many-to-one formula (Theorem 1.1 in Sect. 1.3). Therefore, using (5.18),

$$\mathbf{E}[\zeta_n(z)^2] \leq c_9 (1 + z - K) e^{-z+C}$$

$$+ n^{3/2} e^{-z+C} \sum_{i=1}^{n} \mathbf{E}_{\mathbf{Q}}\left[\mathbf{1}_{\{w_n \in \mathscr{Z}_n^{z,0,K} \cap \mathscr{Y}_n^{z,0,K}\}} \sum_{y \in \mathrm{brot}(w_i)} \varphi_{i,n}(V(y))\right].$$

Assume that we are able to show that for some $c_{11} > 0$ and all sufficiently large n,

$$\sum_{i=1}^{n} \mathbf{E}_{\mathbf{Q}}\left[\mathbf{1}_{\{w_n \in \mathscr{Z}_n^{z,0,K} \cap \mathscr{Y}_n^{z,0,K}\}} \sum_{y \in \mathrm{brot}(w_i)} \varphi_{i,n}(V(y))\right] \leq c_{11} \frac{1+z-K}{n^{3/2}}. \tag{5.23}$$

Then $\mathbf{E}[\zeta_n(z)^2] \leq (c_9 + c_{11})(1 + z - K) e^{C-z}$, for all large n. We already know that $\mathbf{E}[\zeta_n(z)] \geq c_{10} e^{-z}$, which implies $\mathbf{P}(\zeta_n(z) > 0) \geq \frac{\{\mathbf{E}[\zeta_n(z)]\}^2}{\mathbf{E}[\zeta_n(z)^2]} \geq c_{12}(1 + z - K) e^{-z}$, for some constant $c_{12} > 0$. Since $\{\zeta_n(z) > 0\} \subset \{\exists x : |x| = n, \, \min_{1 \leq i \leq n} V(x_i) \geq -z + K, \, \frac{3}{2}\ln n - z \leq V(x) \leq \frac{3}{2}\ln n - z + C\}$, this will complete the proof of Lemma 5.13 by simply taking $K = z$.

It remains to check (5.23). We bound $\varphi_{i,n}(r)$ differently depending on whether $i \leq \lfloor \frac{n}{2} \rfloor + 1$ or $i > \lfloor \frac{n}{2} \rfloor + 1$. In the rest of the proof, we treat $\frac{n}{2}$ as an integer.

First Case: $i \leq \frac{n}{2} + 1$. The $j = 0$ term gives $\varphi_{i,n}(r) = 0$ for $r < -z + K$. For $r \geq -z + K$, we use (5.22) and Lemma A.4 of Appendix A.2 (the probability term in Lemma A.4 being non-decreasing in λ, the lemma applies even if i is close to $\frac{n}{2}$):

$$\varphi_{i,n}(r) \leq n^{3/2} e^{C-z-r} c_{13} \frac{r+z-K+1}{n^{3/2}} = e^C c_{13} e^{-r-z}(r+z-K+1). \tag{5.24}$$

Writing $\mathbf{E_Q}[n, i] := \mathbf{E_Q}[\mathbf{1}_{\{w_n \in \mathscr{L}_n^{z,0,K} \cap \mathscr{Y}_n^{z,0,K}\}} \sum_{y \in \text{brot}(w_i)} \varphi_{i,n}(V(y))]$ and $c_{14} := e^C c_{13}$,

$$\mathbf{E_Q}[n, i] \leq c_{14} \, \mathbf{E_Q}\Big[\mathbf{1}_{\{w_n \in \mathscr{L}_n^{z,0,K} \cap \mathscr{Y}_n^{z,0,K}\}} \sum_{y \in \text{brot}(w_i)} \mathbf{1}_{\{V(y) \geq -z+K\}}$$

$$\times e^{-V(y)-z}(V(y) + z - K + 1)\Big].$$

Obviously, $\mathbf{1}_{\{V(y) \geq -z+K\}}(V(y) + z - K + 1) \leq (V(y) + z - K)_+ + 1$, so we have $\sum_{y \in \text{brot}(w_i)} \mathbf{1}_{\{V(y) \geq -z+K\}} e^{-V(y)-z}(V(y)+z-K+1) \leq \sum_{y \in \text{brot}(w_i)} e^{-V(y)-z}[(V(y)+z-K)_+ + 1]$, which, on $\{w_n \in \mathscr{L}_n^{z,0,K} \cap \mathscr{Y}_n^{z,0,K}\}$, is bounded by $e^{-K} \rho e^{-(i-1)^{1/7}}$ (by definition of $\mathscr{Y}_n^{z,0,K}$ in (5.20)). This yields that

$$\mathbf{E_Q}[n, i] \leq c_{14}\rho e^{-K-(i-1)^{1/7}} \mathbf{Q}(w_n \in \mathscr{L}_n^{z,0,K} \cap \mathscr{Y}_n^{z,0,K})$$

$$\leq \frac{c_{14}\rho c_9(1 + z - K)}{n^{3/2}} e^{-K-(i-1)^{1/7}},$$

by (5.21). As a consequence (and using $-K \leq 0$)

$$\sum_{1 \leq i \leq \frac{n}{2}+1} \mathbf{E_Q}\Big[\mathbf{1}_{\{w_n \in \mathscr{L}_n^{z,0,K} \cap \mathscr{Y}_n^{z,0,K}\}} \sum_{y \in \text{brot}(w_i)} \varphi_{i,n}(V(y))\Big] \leq c_{15} \frac{1 + z - K}{n^{3/2}}. \tag{5.25}$$

Second (and Last) Case: $\frac{n}{2} + 1 < i \leq n$. This time, we bound $\varphi_{i,n}(r)$ slightly differently. Let us go back to (5.22). Since $i > \frac{n}{2} + 1$, we have $a_{j+i}^{(n)} = \frac{3}{2} \ln n - z$ for all $0 \leq j \leq n - i$ (recalling that $L = 0$), thus $\varphi_{i,n}(r) = 0$ for $r < \frac{3}{2} \ln n - z$, whereas for $r \geq \frac{3}{2} \ln n - z$, we have, by Lemma A.1 of Appendix A.2,

$$\varphi_{i,n}(r) \leq n^{3/2} e^{-z+C-r} \frac{c_{16}}{(n - i + 1)^{3/2}} \Big(r - \frac{3}{2} \ln n + z + 1\Big).$$

This is the analogue of (5.24). From here, we can proceed as in the first case: Writing again $\mathbf{E_Q}[n, i] := \mathbf{E_Q}[\mathbf{1}_{\{w_n \in \mathscr{L}_n^{z,0,K} \cap \mathscr{Y}_n^{z,0,K}\}} \sum_{y \in \text{brot}(w_i)} \varphi_{i,n}(V(y))]$ for brevity, we have

$$\mathbf{E_Q}[n, i] \leq \frac{c_{16} e^C n^{3/2}}{(n - i + 1)^{3/2}} \mathbf{E_Q}\Big[\mathbf{1}_{\{w_n \in \mathscr{L}_n^{z,0,K} \cap \mathscr{Y}_n^{z,0,K}\}}$$

$$\times \sum_{y \in \text{brot}(w_i)} e^{-V(y)-z}[(V(y) - \frac{3}{2} \ln n + z)_+ + 1]\Big]$$

$$\leq \frac{c_{16} e^C n^{3/2}}{(n - i + 1)^{3/2}} \frac{\rho e^{-(n-i+1)^{1/7}}}{n^{3/2}} \mathbf{Q}(w_n \in \mathscr{L}_n^{z,0,K} \cap \mathscr{Y}_n^{z,0,K})$$

$$\leq \frac{c_{17}(1 + z - K)}{(n - i + 1)^{3/2} n^{3/2}} e^{-(n-i+1)^{1/7}},$$

where the last inequality comes from (5.21). Consequently,

$$\sum_{\frac{n}{2}+1<i\le n} \mathbf{E}_{\mathbf{Q}}\Big[\mathbf{1}_{\{w_n\in\mathscr{L}_n^{z,0,K}\cap\mathscr{Y}_n^{z,0,K}\}}\sum_{y\in\mathrm{brot}(w_i)}\varphi_{i,\ell}(V(y))\Big]\le c_{18}\,\frac{1+z-K}{n^{3/2}}.$$

Together with (5.25), this yields (5.23). Lemma 5.13 is proved.   □

5.4 Leftmost Position: Limiting Law

Let $(V(x))$ be a branching random walk. We write

$$M_n := \inf_{|x|=n} V(x), \quad n\ge 0.$$

The main result of this section is Theorem 5.15 below, saying that under suitable general assumptions, M_n, centred by a deterministic term, converges weakly to a non-degenerate law.

Recall that $(V(x), |x|=1)$ is distributed according to a point process denoted by $\varXi$. We say that the law of $\varXi$ is **non-lattice** if there exist no $a>0$ and $b\in\mathbb{R}$ such that a.s., $\{V(x), |x|=1\}\subset a\mathbb{Z}+b$.

Theorem 5.15 (Aïdékon [8]) *Under Assumption* (H), *if the distribution of $\varXi$ is non-lattice, then there exists a constant $C^{\min}\in(0,\infty)$ such that for any $u\in\mathbb{R}$,*

$$\lim_{n\to\infty}\mathbf{P}\Big(M_n-\frac{3}{2}\ln n>u\Big)=\mathbf{E}[e^{-C^{\min}e^u D_\infty}],$$

where D_∞ is the almost sure limit of the derivative martingale (D_n).

We recall from Theorem 5.2 (Sect. 5.2) that under the assumptions of Theorem 5.15, $D_\infty>0$ a.s. on the set of non-extinction. In words, Theorem 5.15 tells us that $-M_n+\frac{3}{2}\ln n$ converges weakly to a Gumbel random variable[6] with an independent random shift of size $\ln(C^{\min}D_\infty)$.

While the assumption that the law of $\varXi$ is non-lattice might look somehow peculiar in Theorem 5.15, it is in fact necessary. Without this condition, it is possible to construct an example of branching random walk such that for any deterministic sequence (a_n), M_n-a_n does not converge in distribution; see Lifshits [165].

The rest of the section is devoted to the proof of this deep result. Since the proof is rather technical, we split it into several steps.

---

[6]A (standard) Gumbel random variable ξ has distribution function $\mathbf{P}\{\xi\le u\}=\exp(-e^{-u}),u\in\mathbb{R}$.

5.4.1 Step 1: The Derivative Martingale is Useful

The presence of D_∞ in Theorem 5.15 indicates that the derivative martingale plays a crucial role in the asymptotic behaviour of M_n. The first step in the proof of Theorem 5.15 is to see how the derivative martingale comes into our picture. More concretely, this step says that in the proof of Theorem 5.15, we only need to study the tail probability of $M_n - \frac{3}{2} \ln n$ instead of its weak convergence; our main tool here is the derivative martingale, summing over some conveniently chosen stopping lines.

The basis in the first step in the proof of Theorem 5.15 is the following tail estimate.

Proposition 5.16 (Key Estimate: Tail Estimate for Minimum) *Under Assumption (H), if the distribution of Ξ is non-lattice, we have, for some constant $c_{\text{tail}} \in (0, \infty)$,*

$$\mathbf{P}\left\{M_n - \frac{3}{2}\ln n \leq -z\right\} \sim c_{\text{tail}}\, ze^{-z}, \quad n \to \infty, \ z \to \infty, \tag{5.26}$$

the exact meaning of (5.26) being that

$$\lim_{z\to\infty} \limsup_{n\to\infty} \left| \frac{1}{ze^{-z}} \mathbf{P}\left\{M_n - \frac{3}{2}\ln n \leq -z\right\} - c_{\text{tail}} \right| = 0. \tag{5.27}$$

Let us see why Proposition 5.16 implies Theorem 5.15. Let, as in (5.13),

$$\mathscr{L}[A] := \left\{ x \in \mathbb{T} : V(x) \geq A, \ \max_{0 \leq i < |x|} V(x_i) < A \right\}.$$

For any $u \in \mathbb{R}$, we have

$$\mathbf{P}\left(M_n - \frac{3}{2}\ln n > u\right) \ \text{``} = \text{''} \ \mathbf{E}\left\{ \prod_{x \in \mathscr{L}[A]} [1 - \Phi_{|x|,n}(V(x) - u)] \right\}, \tag{5.28}$$

where

$$\Phi_{k,n}(r) := \mathbf{P}\left\{ M_{n-k} < \frac{3}{2}\ln n - r \right\}, \quad n \geq 1, \ 1 \leq k \leq n, \ r \in \mathbb{R}.$$

By (5.26), $\Phi_{k,n}(z)$ "$\approx$" $c_{\text{tail}}\, ze^{-z}$, so that

$$\mathbf{P}\left(M_n - \frac{3}{2}\ln n > u\right) \ \text{``} \approx \text{''} \ \mathbf{E}\left\{ \prod_{x \in \mathscr{L}[A]} [1 - c_{\text{tail}}\,(V(x) - u)e^{u-V(x)}] \right\}.$$

On the right-hand side, the parameter n disappears, only A stays. We now let $A \to \infty$.

By Remark 5.11 (Sect. 5.2), $\prod_{x \in \mathscr{L}[A]}[1 - c_{\text{tail}}(V(x) - u)e^{u - V(x)}] \to e^{-c_{\text{tail}} e^u D_\infty}$ a.s., so by the dominated convergence theorem, we have $\mathbf{E}\{\prod_{x \in \mathscr{L}[A]}[1 - c_{\text{tail}}(V(x) - u)e^{u - V(x)}]\} \to \mathbf{E}[e^{-c_{\text{tail}} e^u D_\infty}]$, $A \to \infty$. This will yield Theorem 5.15.

The argument presented is only heuristic for the moment. For example, in (5.28), we will run into trouble if $|x| > n$ for some $x \in \mathscr{L}[A]$. Now let us write a rigorous proof.

Proof of Theorem 5.15 Fix $u \in \mathbb{R}$ and $\varepsilon > 0$. By the key estimate (Proposition 5.16), we can choose and fix a sufficiently large A such that

$$\limsup_{n \to \infty} \left| \frac{e^z}{z} \mathbf{P}\left\{ M_n \leq \frac{3}{2} \ln n - z \right\} - c_{\text{tail}} \right| \leq \frac{\varepsilon}{2}, \quad \forall z \geq A - u. \tag{5.29}$$

We also fix $A_0 = A_0(A, \varepsilon) > A$ sufficiently large such that $\mathbf{P}(\mathscr{Y}_{A,A_0}) > 1 - \varepsilon$, where

$$\mathscr{Y}_{A,A_0} := \left\{ \sup_{x \in \mathscr{L}[A]} |x| \leq A_0, \ \sup_{x \in \mathscr{L}[A]} V(x) \leq A_0 \right\}.$$

Instead of having an identity in (5.28), we use an upper bound and a lower bound, both of which are rigorous: writing $\overline{F}_n(u) := \mathbf{P}(M_n - \frac{3}{2} \ln n > u)$, then for all $n > A_0$,

$$\overline{F}_n(u) \leq \mathbf{E}\left\{ \mathbf{1}_{\mathscr{Y}_{A,A_0}} \prod_{x \in \mathscr{L}[A]} [1 - \Phi_{|x|,n}(V(x) - u)] \right\} + \varepsilon,$$

$$\overline{F}_n(u) \geq \mathbf{E}\left\{ \mathbf{1}_{\mathscr{Y}_{A,A_0}} \prod_{x \in \mathscr{L}[A]} [1 - \Phi_{|x|,n}(V(x) - u)] \right\}.$$

By (5.29), there exists a sufficiently large integer $n_0 \geq 1$ such that

$$\left| \frac{e^z}{z} \Phi_{k,n}(z) - c_{\text{tail}} \right| \leq \varepsilon, \quad \forall n \geq n_0, \ 0 \leq k \leq A_0, \ z \in [A - u, A_0 - u].$$

Therefore,

$$\overline{F}_n(u) \leq \mathbf{E}\left\{ \mathbf{1}_{\mathscr{Y}_{A,A_0}} \prod_{x \in \mathscr{L}[A]} [1 - (c_{\text{tail}} - \varepsilon)(V(x) - u)e^{-[V(x) - u)]}] \right\} + \varepsilon$$

$$\leq \mathbf{E}\left\{ \prod_{x \in \mathscr{L}[A]} [1 - (c_{\text{tail}} - \varepsilon)(V(x) - u)e^{-[V(x) - u)]}] \right\} + \varepsilon,$$

and, similarly,

$$\overline{F}_n(u) \geq \mathbf{E}\Big\{ 1_{\mathscr{Y}_{A,A_0}} \prod_{x \in \mathscr{Z}[A]} [1 - (c_{\text{tail}} + \varepsilon)(V(x) - u)e^{-[V(x)-u)]}] \Big\} - \varepsilon$$

$$\geq \mathbf{E}\Big\{ \prod_{x \in \mathscr{Z}[A]} [1 - (c_{\text{tail}} + \varepsilon)(V(x) - u)e^{-[V(x)-u)]}] \Big\} - 2\varepsilon,$$

where, in the last inequality, we use the fact that $\mathbf{P}(\mathscr{Y}_{A,A_0}) \geq 1 - \varepsilon$. Letting $n \to \infty$ gives that

$$\limsup_{n \to \infty} \overline{F}_n(u) \leq \mathbf{E}\Big\{ \prod_{x \in \mathscr{Z}[A]} [1 - (c_{\text{tail}} - \varepsilon)(V(x) - u)e^{-[V(x)-u)]}] \Big\} + \varepsilon,$$

$$\liminf_{n \to \infty} \overline{F}_n(u) \geq \mathbf{E}\Big\{ \prod_{x \in \mathscr{Z}[A]} [1 - (c_{\text{tail}} + \varepsilon)(V(x) - u)e^{-[V(x)-u)]}] \Big\} - 2\varepsilon.$$

On the right-hand sides, we let $A \to \infty$. We have already noted that $\mathbf{E}\{\prod_{x \in \mathscr{Z}[A]}[1 - c(V(x) - u)e^{u-V(x)}]\} \to \mathbf{E}[e^{-ce^u D_\infty}]$ for any given constant $c > 0$, hence

$$\limsup_{n \to \infty} \overline{F}_n(u) \leq \mathbf{E}[e^{-(c_{\text{tail}}-\varepsilon)e^u D_\infty}] + \varepsilon,$$

$$\liminf_{n \to \infty} \overline{F}_n(u) \geq \mathbf{E}[e^{-(c_{\text{tail}}+\varepsilon)e^u D_\infty}] - 2\varepsilon.$$

As $\varepsilon > 0$ can be arbitrarily small, the proof of Theorem 5.15 will be complete once Proposition 5.16 is established. □

Remark 5.17 In this step, we did not use the assumption of non-lattice of the distribution of $\varXi$ (which was, however, necessary for the validity of the key estimate). □

5.4.2 Step 2: Proof of the Key Estimate

Our aim is now to prove the key estimate (Proposition 5.16), of which we recall the statement: for some $0 < c_{\text{tail}} < \infty$,

$$\mathbf{P}\Big\{ M_n - \frac{3}{2}\ln n \leq -z \Big\} \sim c_{\text{tail}} z e^{-z}, \qquad n \to \infty, z \to \infty. \qquad ((5.26))$$

We prove this with the aid of several technical estimates, whose proofs (and sometimes, statements) are postponed to the forthcoming subsections for the sake of the clarity. Let $L \geq 0$ and $0 \leq K \leq z \leq \frac{3}{2}\ln n - L$. Recall $(a_i^{(n)}, 0 \leq i \leq n)$ from (5.16):

$$a_i^{(n)} = a_i^{(n)}(z, L, K) := \begin{cases} -z + K & \text{if } 0 \leq i \leq \lfloor\frac{n}{2}\rfloor, \\ \frac{3}{2}\ln n - z - L & \text{if } \lfloor\frac{n}{2}\rfloor < i \leq n. \end{cases}$$

Consider, for all $x \in \mathbb{T}$ with $|x| = n$,

$$E_n(x) := \left\{ V(x_i) \geq a_i^{(n)}, \forall 0 \leq i \leq n, V(x) \leq \frac{3}{2}\ln n - z \right\}. \tag{5.30}$$

In order not to burden our notation, we do not write explicitly the dependence in (z, K, L) of the event E_n.

Let $m^{(n)}$ be a vertex chosen uniformly in the set $\{x \in \mathbb{T} : |x| = n, V(x) = M_n\}$, which is the set of particles achieving the minimum M_n. So $m^{(n)}$ is well-defined as long as the system survives until generation n.

Our first preliminary result is as follows.

Lemma 5.18 *For any $\varepsilon > 0$, there exist $L_0 > 0$ and $n_0 \geq 2$ such that for all $L \geq L_0$, $n \geq n_0$ and all $0 \leq K \leq z \leq \frac{3}{2}\ln n - L$, we have*

$$\mathbf{P}\left(M_n - \frac{3}{2}\ln n \leq -z, E_n(m^{(n)})^c\right) \leq e^{K-z} + \varepsilon(1 + z - K)e^{-z}, \tag{5.31}$$

and moreover,

$$\mathbf{P}\left(M_n - \frac{3}{2}\ln n \leq -z, \min_{0 \leq i \leq \frac{n}{2}} V(m_i^{(n)}) \geq -z + K, E_n(m^{(n)})^c\right)$$
$$\leq \varepsilon(1 + z - K)e^{-z}, \tag{5.32}$$

where $m_i^{(n)}$ is the ancestor of $m^{(n)}$ in the i-th generation.

Since both e^{K-z} and $\varepsilon(1 + z - K)e^{-z}$ can be much smaller than ze^{-z} (for $n \to \infty$ and then $z \to \infty$, as long as K is a fixed constant, and $\varepsilon > 0$ very small), Lemma 5.18 tells us that in the study of $\mathbf{P}(M_n - \frac{3}{2}\ln n \leq -z)$, we can limit ourselves to the event $E_n(m^{(n)})$.

Lemma 5.18 is proved in Sect. 5.4.3.

We start with our proof of the key estimate, Proposition 5.16. Let $\varepsilon > 0$. By inequality (5.31) in Lemma 5.18,

$$\left|\mathbf{P}\left(M_n - \frac{3}{2}\ln n \leq -z\right) - \Phi_n^{E_n}(z)\right| \leq e^{K-z} + \varepsilon(1 + z - K)e^{-z}, \tag{5.33}$$

where

$$\Phi_n^{E_n}(z) := \mathbf{P}\Big(E_n(m^{(n)})\Big) = \mathbf{P}\Big(M_n - \frac{3}{2}\ln n \le -z, \; E_n(m^{(n)})\Big),$$

the second identity being a consequence of the obvious fact that $E_n(x) \subset \{M_n - \frac{3}{2}\ln n \le -z\}$ for any $x \in \mathbb{T}$ with $|x| = n$. For further use, we note that if instead of (5.31), we use inequality (5.32) in Lemma 5.18, then we also have

$$\left| \mathbf{P}\Big(M_n - \frac{3}{2}\ln n \le -z, \; \min_{0 \le i \le \frac{n}{2}} V(m_i^{(n)}) \ge -z + K\Big) - \Phi_n^{E_n}(z) \right|$$

$$\le \varepsilon(1 + z - K)e^{-z}. \tag{5.34}$$

We now study $\Phi_n^{E_n}(z)$. Since $m^{(n)}$ is uniformly chosen among vertices in generation n realizing the minimum, we have

$$\Phi_n^{E_n}(z) = \mathbf{E}\Big[\sum_{x \in \mathbb{T}: |x|=n} \mathbf{1}_{\{m^{(n)}=x\}} \mathbf{1}_{E_n(x)} \Big]$$

$$= \mathbf{E}\Big[\frac{\sum_{x \in \mathbb{T}: |x|=n} \mathbf{1}_{\{V(x)=M_n\}} \mathbf{1}_{E_n(x)}}{\sum_{|x|=n} \mathbf{1}_{\{V(x)=M_n\}}} \Big]. \tag{5.35}$$

We now use the spinal decomposition theorem as in Example 4.5 of Sect. 4.6: $\mathbf{Q}_a(A) = \mathbf{E}_a[\frac{W_n}{e^{-a}} \mathbf{1}_A]$ for $A \in \mathscr{F}_n$ and $n \ge 1$; $\mathbf{Q}_a(w_n = x \mid \mathscr{F}_n) = \frac{e^{-V(x)}}{W_n}$ for any n and any vertex $x \in \mathbb{T}$ such that $|x| = n$; $(V(w_n) - V(w_{n-1}), n \ge 1)$, is, under $\mathbf{Q}_a$, a sequence of i.i.d. random variables whose common distribution is that of S_1 under $\mathbf{P}_0$. Accordingly, working under $\mathbf{Q} = \mathbf{Q}_0$,

$$\Phi_n^{E_n}(z) = \mathbf{E}_{\mathbf{Q}}\Big[\frac{\sum_{x \in \mathbb{T}: |x|=n} \frac{1}{W_n} \mathbf{1}_{\{V(x)=M_n\}} \mathbf{1}_{E_n(x)}}{\sum_{x \in \mathbb{T}: |x|=n} \mathbf{1}_{\{V(x)=M_n\}}} \Big] \quad \text{(def. of } \mathbf{Q})$$

$$= \mathbf{E}_{\mathbf{Q}}\Big[\frac{\sum_{x \in \mathbb{T}: |x|=n} e^{V(x)} \mathbf{1}_{\{w_n=x\}} \mathbf{1}_{\{V(x)=M_n\}} \mathbf{1}_{E_n(x)}}{\sum_{x \in \mathbb{T}: |x|=n} \mathbf{1}_{\{V(x)=M_n\}}} \Big],$$

where the second equality follows from $\mathbf{Q}(w_n = x \mid \mathscr{F}_n) = \frac{e^{-V(x)}}{W_n}$ (see (4.16)). This leads to:

$$\Phi_n^{E_n}(z) = \mathbf{E}_{\mathbf{Q}}\Big[\frac{e^{V(w_n)} \mathbf{1}_{\{V(w_n)=M_n\}}}{\sum_{x \in \mathbb{T}: |x|=n} \mathbf{1}_{\{V(x)=M_n\}}} \mathbf{1}_{E_n(w_n)} \Big].$$

At this stage, it is convenient to introduce another event $\mathscr{E}_{n,b} = \mathscr{E}_{n,b}(z)$, and claim that in the $\mathbf{Q}$-expectation expression, we can integrate only on the $\mathscr{E}_{n,b}$: let $b \ge 1$ be an integer, and let

$$
\mathscr{E}_{n,b} := \bigcap_{i=0}^{n-b} \left\{ \min_{x \in \mathbb{T}: \, |x|=n, \, x>y} V(x) > \frac{3}{2} \ln n - z, \ \forall y \in \mathrm{brot}(w_i) \right\}, \tag{5.36}
$$

where $\mathrm{brot}(w_i)$ denotes as before the set of brothers of w_i. Since $\mathscr{E}_{n,b}$ involves w_n, it is well-defined only under $\mathbf{Q}$. Another simple observation is that on $\mathscr{E}_{n,b} \cap \{M_n \leq \frac{3}{2} \ln n - z\}$, any particle at the leftmost position at generation n must be separated from the spine after generation $n-b$.

Here is our second preliminary result.

Lemma 5.19 *For any $\eta > 0$ and $L \geq 0$, there exist $K_0 > 0$, $b_0 \geq 1$ and $n_0 \geq 2$ such that for all $n \geq n_0$, $b_0 \leq b < n$ and $K_0 \leq K \leq z \leq \frac{3}{2} \ln n - L$,*

$$
\mathbf{Q}\Big(E_n(w_n)\backslash\mathscr{E}_{n,b}\Big) \leq \frac{\eta\,(1+z-K)}{n^{3/2}}.
$$

Lemma 5.19 is proved in Sect. 5.4.4.

We continue with our proof of the key estimate (Proposition 5.16). Write

$$
\Phi_n^{E_n, \, \mathscr{E}_{n,b}}(z) := \mathbf{E}_{\mathbf{Q}}\left[\frac{e^{V(w_n)} \, \mathbf{1}_{\{V(w_n)=M_n\}}}{\sum_{x \in \mathbb{T}: \, |x|=n} \mathbf{1}_{\{V(x)=M_n\}}} \, \mathbf{1}_{E_n(w_n) \cap \mathscr{E}_{n,b}} \right].
$$

[It is the same $\mathbf{Q}$-integration as for $\Phi_n^{E_n}(z)$ except that we integrate also on $\mathscr{E}_{n,b}$.] By definition,

$$
0 \leq \Phi_n^{E_n}(z) - \Phi_n^{E_n, \, \mathscr{E}_{n,b}}(z) \leq \mathbf{E}_{\mathbf{Q}}\left[e^{V(w_n)} \, \mathbf{1}_{E_n(w_n) \backslash \mathscr{E}_{n,b}} \right] \leq e^{\frac{3}{2}\ln n - z} \, \mathbf{Q}\Big(E_n(w_n)\backslash\mathscr{E}_{n,b}\Big).
$$

By Lemma 5.19, there exist $K_0 > 0$, $b_0 \geq 1$ and $n_0 \geq 2$ such that for $b \geq b_0$, $n \geq n_0$ and $z \geq K \geq K_0$,

$$
0 \leq \Phi_n^{E_n}(z) - \Phi_n^{E_n, \, \mathscr{E}_{n,b}}(z) \leq \eta\,(1+z-K)\,e^{-z}. \tag{5.37}
$$

Let us look at $\Phi_n^{E_n, \, \mathscr{E}_{n,b}}(z)$. On the event $E_n(w_n)$, we have $M_n \leq \frac{3}{2} \ln n - z$, so on the event $E_n(w_n) \cap \mathscr{E}_{n,b}$, any particle at the leftmost position at generation n must be a descendant of w_{n-b}. Hence, on $E_n(w_n) \cap \mathscr{E}_{n,b}$, $\sum_{x \in \mathbb{T}: \, |x|=n} \mathbf{1}_{\{V(x)=M_n\}} = \sum_{x \in \mathbb{T}: \, |x|=n, \, x>w_{n-b}} \mathbf{1}_{\{V(x)=M_n\}}$ (this is why $\mathscr{E}_{n,b}$ was introduced). As such,

$$
\Phi_n^{E_n, \, \mathscr{E}_{n,b}}(z) = \mathbf{E}_{\mathbf{Q}}\left[\frac{e^{V(w_n)} \, \mathbf{1}_{\{V(w_n)=M_n\}}}{\sum_{x \in \mathbb{T}: \, |x|=n, \, x>w_{n-b}} \mathbf{1}_{\{V(x)=M_n\}}} \, \mathbf{1}_{E_n(w_n) \cap \mathscr{E}_{n,b}} \right].
$$

We use the spinal decomposition theorem in Example 4.5 of Sect. 4.6: by applying the branching property at the vertex w_{n-b}, we have, for $n > 2b$ (so that $\frac{n}{2} < n - b$),

$$\Phi_n^{E_n, \mathscr{E}_{n,b}}(z) = \mathbf{E}_{\mathbf{Q}}\left[\widehat{F}_{L,b}(V(w_{n-b}))\, \mathbf{1}_{\{V(w_i)\geq a_i^{(n)},\ \forall 0\leq i\leq n-b\}}\, \mathbf{1}_{\mathscr{E}_{n,b}}\right],$$

where, for $u \geq a_{n-b}^{(n)} := \frac{3}{2}\ln n - z - L$,

$$\widehat{F}_{L,b}(u) := \mathbf{E}_{\mathbf{Q}_u}\left[\frac{e^{V(w_b)}\, \mathbf{1}_{\{V(w_b)=M_b\}}}{\sum_{x\in\mathbb{T}:\, |x|=b}\mathbf{1}_{\{V(x)=M_b\}}}\, \mathbf{1}_{\{\min_{0\leq j\leq b} V(w_j)\geq a_{n-b}^{(n)},\ V(w_b)-\frac{3}{2}\ln n\leq -z\}}\right].$$

We do not need the event $\mathscr{E}_{n,b}$ any more (which served only to take out the denominator $\sum_{x\in\mathbb{T}:\, |x|=n,\, x>w_{n-b}}\mathbf{1}_{\{V(x)=M_n\}}$), so let us get rid of it by introducing

$$\hat{\Phi}_n^{E_n}(z) := \mathbf{E}_{\mathbf{Q}}\left[\widehat{F}_{L,b}(V(w_{n-b}))\, \mathbf{1}_{\{V(w_i)\geq a_i^{(n)},\ \forall 0\leq i\leq n-b\}}\right],$$

which is the same $\mathbf{Q}$-expectation as for $\Phi_n^{E_n, \mathscr{E}_{n,b}}(z)$, but without the factor $\mathbf{1}_{\mathscr{E}_{n,b}}$. We have

$$\hat{\Phi}_n^{E_n}(z) - \Phi_n^{E_n, \mathscr{E}_{n,b}}(z) = \mathbf{E}_{\mathbf{Q}}\left[\mathbf{1}_{\{V(w_i)\geq a_i^{(n)},\ \forall 0\leq i\leq n-b\}}\, \mathbf{1}_{\mathscr{E}_{n,b}^c}\, \widehat{F}_{L,b}(V(w_{n-b}))\right].$$

By definition, $\widehat{F}_{L,b}(u)$ is bounded by $e^{\frac{3}{2}\ln n - z}\, \mathbf{Q}_u(\min_{0\leq j\leq b} V(w_j) \geq \frac{3}{2}\ln n - z - L,\ V(w_b) - \frac{3}{2}\ln n \leq -z)$, so that

$$\hat{\Phi}_n^{E_n}(z) - \Phi_n^{E_n, \mathscr{E}_{n,b}}(z) \leq e^{\frac{3}{2}\ln n - z}\, \mathbf{E}_{\mathbf{Q}}\left[\mathbf{1}_{\{V(w_i)\geq a_i^{(n)},\ \forall 0\leq i\leq n-b\}}\, \mathbf{1}_{\mathscr{E}_{n,b}^c}\, \mathbf{Q}_{V(w_{n-b})}\right.$$
$$\left. \times \left(\min_{0\leq j\leq b} V(w_j) \geq \frac{3}{2}\ln n - z - L,\ V(w_b) - \frac{3}{2}\ln n \leq -z\right)\right],$$

which, by the branching property at the vertex w_{n-b} again, yields

$$\hat{\Phi}_n^{E_n}(z) - \Phi_n^{E_n, \mathscr{E}_{n,b}}(z) \leq e^{\frac{3}{2}\ln n - z}\, \mathbf{Q}\left(E_n(w_n)\backslash\mathscr{E}_{n,b}\right).$$

By Lemma 5.19, this yields (for $b \geq b_0$, $n \geq n_0$ and $K_0 \leq K \leq z \leq \frac{3}{2}\ln n - L$)

$$0 \leq \hat{\Phi}_n^{E_n}(z) - \Phi_n^{E_n, \mathscr{E}_{n,b}}(z) \leq \eta\,(1 + z - K)\, e^{-z}. \tag{5.38}$$

We now study $\hat{\Phi}_n^{E_n}(z)$. It is more convenient to make a transformation of the function $\widehat{F}_{L,b}$ by setting, for $v \geq 0$,

$$F_{L,b}(v) := n^{-3/2}\, e^{z+L}\, \widehat{F}_{L,b}\left(v + \frac{3}{2}\ln n - z - L\right)$$

$$= \mathbf{E}_{\mathbf{Q}_v}\left[\frac{e^{V(w_b)}\,\mathbf{1}_{\{V(w_b)=M_b\}}}{\sum_{x\in\mathbb{T}:\,|x|=b}\mathbf{1}_{\{V(x)=M_b\}}}\,\mathbf{1}_{\{\min_{0\le j\le b} V(w_j)\ge 0,\,V(w_b)\le L\}}\right].$$

Then

$$\hat{\phi}_n^{E_n}(z) = n^{3/2}\,e^{-z-L}\,\mathbf{E}_{\mathbf{Q}}\left[F_{L,b}(V(w_{n-b})-\frac{3}{2}\ln n + z + L)\,\mathbf{1}_{\{V(w_i)\ge a_i^{(n)},\,\forall 0\le i\le n-b\}}\right].$$

By the spinal decomposition theorem, $(V(w_i))$ under $\mathbf{Q}$ is a centred random walk. We apply Proposition 5.22 (see Sect. 5.4.5 below) to the function $F_{L,b}$ (which is easily checked to satisfy the assumptions in Proposition 5.22) and in its notation:

$$\lim_{n\to\infty}\hat{\phi}_n^{E_n}(z) = C_{L,b}\,e^{-z}\,R(z-K),$$

where $C_{L,b} := e^{-L}\,C_+C_-(\frac{\pi}{2\sigma^2})^{1/2}\int_0^\infty F_{L,b}(u)R_-(u)\,du \in (0,\infty)$. Combining this with (5.33), (5.37) and (5.38) gives that, for $L \ge L_0$, $b \ge b_0$ and $K_0 \le K \le z \le \frac{3}{2}\ln n - L$,

$$\limsup_{n\to\infty}\left|\mathbf{P}\left(M_n - \frac{3}{2}\ln n \le -z\right) - C_{L,b}\,e^{-z}\,R(z-K)\right|$$

$$\le e^{K-z} + \varepsilon(1 + z - K)e^{-z} + 2\eta(1 + z - K)\,e^{-z}.$$

Recall that $\lim_{z\to\infty}\frac{R(z-K)}{z} = c_{\mathrm{ren}} \in (0,\infty)$ (see (A.4) and (A.5) in Appendix A.1). This yields that for $L \ge L_0$ and $b \ge b_0 = b_0(\eta, L)$,

$$\limsup_{z\to\infty}\limsup_{n\to\infty}\left|\frac{\mathbf{P}(M_n - \frac{3}{2}\ln n \le -z)}{z\,e^{-z}} - C_{L,b}\,c_{\mathrm{ren}}\right| \le \varepsilon + 2\eta.$$

In particular,

$$\limsup_{z\to\infty}\limsup_{n\to\infty}\frac{\mathbf{P}(M_n - \frac{3}{2}\ln n \le -z)}{z\,e^{-z}} < \infty. \qquad (5.39)$$

It remains to prove that $\lim_{L\to\infty}\lim_{b\to\infty} C_{L,b}$ exists, and lies in $(0,\infty)$. Consider

$$\underline{d}_{L,b} := \liminf_{z\to\infty}\liminf_{n\to\infty}\frac{\Phi_n^{E_n,\mathscr{E}_{n,b}}(z)}{z\,e^{-z}}, \qquad \overline{d}_{L,b} := \limsup_{z\to\infty}\limsup_{n\to\infty}\frac{\Phi_n^{E_n,\mathscr{E}_{n,b}}(z)}{z\,e^{-z}}.$$

We have proved that for $L \ge L_0$, $\eta > 0$ and all $b \ge b_0(\eta, L)$,

$$C_{L,b}\,c_{\mathrm{ren}} - \eta \le \underline{d}_{L,b} \le \overline{d}_{L,b} \le C_{L,b}\,c_{\mathrm{ren}}.$$

We let $b \to \infty$. Since $\Phi_n^{E_n, \mathscr{E}_{n,b}}(z)$ is non-decreasing in b, we can define

$$\underline{d}_L := \lim_{b \to \infty} \underline{d}_{L,b}, \qquad \overline{d}_L := \lim_{b \to \infty} \overline{d}_{L,b},$$

to see that

$$\limsup_{b \to \infty} C_{L,b} \le \frac{\underline{d}_L}{c_{\mathrm{ren}}} \le \frac{\overline{d}_L}{c_{\mathrm{ren}}} \le \liminf_{b \to \infty} C_{L,b}.$$

This implies that for $L \ge L_0$,

$$\lim_{b \to \infty} C_{L,b} = \frac{\underline{d}_L}{c_{\mathrm{ren}}} = \frac{\overline{d}_L}{c_{\mathrm{ren}}}.$$

Since $\Phi_n^{E_n, \mathscr{E}_{n,b}}(z)$ is also non-decreasing in L, the limit $\lim_{L \to \infty} \underline{d}_L$ exists as well, which yields the existence of the limit

$$\tilde{c}_{\mathrm{tail}} := \lim_{L \to \infty} \lim_{b \to \infty} C_{L,b}.$$

Moreover, (5.39) implies $\widetilde{c}_{\mathrm{tail}} < \infty$.

It is also easy to see that $\widetilde{c}_{\mathrm{tail}} > 0$. Indeed, by (5.34), and taking $K = z$, we get for $L \ge L_0$ and $b \ge b_0 = b_0(\eta, L)$,

$$\limsup_{z \to \infty} \limsup_{n \to \infty} \left| \frac{\mathbf{P}(M_n - \frac{3}{2} \ln n \le -z, \ \min_{0 \le i \le \frac{n}{2}} V(m_i^{(n)}) \ge 0)}{\mathrm{e}^{-z}} - C_{L,b}\, c_{\mathrm{ren}} \right| \le \varepsilon + 2\eta.$$

[Note that the normalising function becomes e^{-z}, instead of $z\,\mathrm{e}^{-z}$.] And we know that $\tilde{c}_{\mathrm{tail}} = \lim_{L \to \infty} \lim_{b \to \infty} C_{L,b}$ exists. By Lemma 5.13, there exists $c_{19} > 0$, independent of L and b, such that

$$\limsup_{z \to \infty} \limsup_{n \to \infty} \frac{\mathbf{P}(M_n - \frac{3}{2} \ln n \le -z, \ \min_{0 \le i \le \frac{n}{2}} V(m_i^{(n)}) \ge 0)}{\mathrm{e}^{-z}} \ge c_{19},$$

which yields $\widetilde{c}_{\mathrm{tail}} \ge \frac{c_{19}}{c_{\mathrm{ren}}} > 0$.

Proposition 5.16 is proved with $c_{\mathrm{tail}} := \widetilde{c}_{\mathrm{tail}}\, c_{\mathrm{ren}} \in (0, \infty)$. $\qquad \square$

5.4.3 Step 3a: Proof of Lemma 5.18

Before proceeding to the proof of Lemma 5.18, let us prove a preliminary estimate.

Lemma 5.20 *There exist constants $c_{20} > 0$ and $c_{21} > 0$ such that for all sufficiently large n, all $L \geq 0$, $u \geq 0$ and $z \geq 0$,*

$$\mathbf{P}_u\left(\exists x \in \mathbb{T}:\ |x| = n,\ \min_{1 \leq i \leq n} V(x_i) \geq 0,\ V(x) \simeq \frac{3}{2}\ln n - z, \right.$$

$$\left. \min_{\frac{n}{2} \leq j \leq n} V(x_j) \simeq \frac{3}{2}\ln n - z - L\right) \leq c_{20}\,(1 + u)\mathrm{e}^{-c_{21}\,L - u - z}, \qquad (5.40)$$

where $a \simeq b$ is short for $|a - b| \leq 1$.

Proof of Lemma 5.20 There is nothing to prove if $\frac{3}{2}\ln n - z - L < -1$; so we assume $\frac{3}{2}\ln n - z - L \geq -1$.

For brevity, we write $p_{(5.40)}$ for the probability expression on the left-hand side of (5.40), and $\alpha = \alpha(n, z, L) := \frac{3}{2}\ln n - z - L \geq -1$. Let $\ell \in [\frac{3}{4}n, n)$ be an integer. We observe that

$$\{\exists x \in \mathbb{T}:\ |x| = n,\ \min_{\frac{n}{2} \leq j \leq n} V(x_j) \simeq \alpha\} \subset \bigcup_{j=\frac{n}{2}}^{\ell}\{\exists x \in \mathbb{T}:\ |x| = n,\ V(x_j) \simeq \alpha\}$$

$$\cup \bigcup_{j=\ell+1}^{n}\{\exists x \in \mathbb{T}:\ |x| = j,\ V(x_j) \simeq \alpha\}.$$

[Note that in the last event, $|x|$ becomes j, not n any more: this trick goes back at least to Kesten [154].] By the many-to-one formula (Theorem 1.1 in Sect. 1.3), and in its notation,

$$p_{(5.40)} \leq \sum_{j=\frac{n}{2}}^{\ell} \mathbf{E}_u\left(\mathrm{e}^{S_n - u}\mathbf{1}_{\{\min_{1 \leq i \leq \frac{n}{2}} S_i \geq 0,\ S_n \simeq \frac{3}{2}\ln n - z,\ S_j \simeq \alpha,\ \min_{\frac{n}{2} \leq k \leq n} S_k \geq \alpha - 1\}}\right)$$

$$+ \sum_{j=\ell+1}^{n} \mathbf{E}_u\left(\mathrm{e}^{S_j - u}\mathbf{1}_{\{\min_{1 \leq i \leq \frac{n}{2}} S_i \geq 0,\ S_j \simeq \alpha,\ \min_{\frac{n}{2} \leq k \leq j} S_k \geq \alpha - 1\}}\right)$$

$$\leq \mathrm{e}^{\frac{3}{2}\ln n - z + 1 - u}\sum_{j=\frac{n}{2}}^{\ell}\mathbf{P}_u(E_j^{(n)}) + \mathrm{e}^{\alpha + 1 - u}\sum_{j=\ell+1}^{n}\mathbf{P}_u(\widetilde{E}_j^{(n)}),$$

where

$$E_j^{(n)} := \left\{\min_{1 \leq i \leq \frac{n}{2}} S_i \geq 0,\ \min_{\frac{n}{2} \leq k \leq n} S_k \geq \alpha - 1,\ S_j \simeq \alpha,\ S_n \simeq \frac{3}{2}\ln n - z\right\},$$

$$\widetilde{E}_j^{(n)} := \left\{\min_{1 \leq i \leq \frac{n}{2}} S_i \geq 0,\ \min_{\frac{n}{2} \leq k \leq j} S_k \geq \alpha - 1,\ S_j \simeq \alpha\right\}.$$

Let $j \in [\frac{n}{2}, n)$. By the Markov property at time j,

$$\mathbf{P}_u(E_j^{(n)}) \leq \mathbf{P}_u(\widetilde{E}_j^{(n)}) \times \mathbf{P}\Big(\min_{1 \leq i \leq n-j} S_i \geq -2, \ |S_{n-j} - (\tfrac{3}{2} \ln n - z - \alpha)| \leq 2 \Big).$$

[Recall that $\alpha = \frac{3}{2} \ln n - z - L$ by our notation, so $\frac{3}{2} \ln n - z - \alpha$ is simply L, whereas $e^{\alpha+1-u} = n^{3/2} e^{-z-L+1-u}$.] The last probability expression $\mathbf{P}(\cdots)$ on the right-hand side is, according to Lemma A.2 (see Appendix A.2), bounded by $c_{22} \frac{L+5}{(n-j)^{3/2}}$. So

$$p_{(5.40)} \leq c_{23}\, n^{3/2}\, e^{-z-u} \Big[\sum_{j=\frac{n}{2}}^{\ell} \frac{L+5}{(n-j)^{3/2}} \mathbf{P}_u(\widetilde{E}_j^{(n)}) + e^{-L} \sum_{j=\ell+1}^{n} \mathbf{P}_u(\widetilde{E}_j^{(n)}) \Big].$$

We still need to study $\mathbf{P}_u(\widetilde{E}_j^{(n)})$ for $j \in [\frac{n}{2}, n)$. In order to cover also the case $-1 \leq \alpha \leq 1$, we first write the trivial inequality

$$\mathbf{P}_u(\widetilde{E}_j^{(n)}) \leq \mathbf{P}\Big(\min_{1 \leq i \leq \frac{n}{2}} S_i \geq -u - 2, \ \min_{\frac{n}{2} \leq k \leq j} S_k \geq \alpha - u - 1, \ S_j \simeq \alpha - u \Big).$$

[It is an inequality because, instead of $\min_{1 \leq i \leq \frac{n}{2}} S_i \geq -u$, we write $\min_{1 \leq i \leq \frac{n}{2}} S_i \geq -u - 2$.] We use two different strategies depending on the value of j: if $j \in [\frac{3}{4}n, n]$, then we are entitled[7] to use Lemma A.4 of Appendix A.2, to see that $\mathbf{P}_u(\widetilde{E}_j^{(n)}) \leq c_{24} \frac{u+3}{j^{3/2}}$, which is bounded by $c_{24} \frac{u+3}{n^{3/2}}$ because $j \geq \frac{n}{2}$; if $j \in [\frac{n}{2}, \frac{3}{4}n)$, we simply say that (recalling that $\alpha \geq -1$) $\mathbf{P}_u(\widetilde{E}_j^{(n)}) \leq \mathbf{P}(\min_{1 \leq i \leq j} S_i \geq -u - 2, \ S_j \simeq \alpha - u)$, which, by Lemma A.2 again, is bounded by $c_{25} \frac{(u+3)(\alpha+4)}{j^{3/2}}$. Consequently,

$$p_{(5.40)} \leq c_{23}\, n^{3/2}\, e^{-z-u} \Big[\Big(\sum_{j=\frac{n}{2}}^{\frac{3}{4}n-1} \frac{L+5}{(n-j)^{3/2}} c_{25} \frac{(u+3)(\alpha+4)}{j^{3/2}} +$$

$$+ \sum_{j=\frac{3}{4}n}^{\ell} \frac{L+5}{(n-j)^{3/2}} c_{24} \frac{u+3}{n^{3/2}} \Big) + e^{-L} \sum_{j=\ell+1}^{n} c_{24} \frac{u+3}{n^{3/2}} \Big]$$

$$\leq c_{26}\, (u+3) e^{-z-u} \Big[(L+5)\Big(\frac{\alpha+4}{n^{1/2}} + \frac{1}{(n-\ell)^{1/2}} \Big) + e^{-L} (n-\ell) \Big].$$

Recall that $\alpha = \frac{3}{2} \ln n - z - L \leq \frac{3}{2} \ln n$; so $\alpha + 4 \leq 9 \ln n$. Recall also that $L \leq \frac{3}{2} \ln n + 1$ by assumption. Since the inequality we have just proved holds uniformly

---

[7]This is why we assume $j \geq \frac{3}{4}n$; otherwise, we would not be able to apply this lemma when j is close to $\frac{n}{2}$.

in $\ell \in [\frac{3}{4}n, n)$, we choose $\ell := n - \lfloor c_{27} e^{c_{28} L} \rfloor$, where c_{27} and c_{28} are sufficiently small constants, to conclude. $\qquad\square$

Remark 5.21 Applying Lemma 5.20 to $z+k$ (with $k \geq 0$) instead of z, and summing over integer values of k and L gives that for some constant $c_{29} > 0$, all sufficiently large n, and all $u \geq 0$ and $z \geq 0$,

$$\mathbf{P}_u\left(\exists x : |x| = n, \ \min_{1 \leq i \leq n} V(x_i) \geq 0, \ V(x) \leq \frac{3}{2}\ln n - z\right) \leq c_{29}\,(1 + u)e^{-u-z}.$$
$$(5.41)$$

On the other hand, for any $r > 0$,

$$\mathbf{P}\left(\inf_{y \in \mathbb{T}} V(y) \leq -r\right) \leq \sum_{j=1}^{\infty} \mathbf{E}\left[\sum_{|y|=j} \mathbf{1}_{\{V(y) \leq -r, \ \min_{i: \, 0 \leq i < j} V(y_i) > -r\}}\right].$$

By the many-to-one formula (Theorem 1.1 in Sect. 1.3), the right-hand side is

$$= \sum_{j=1}^{\infty} \mathbf{E}\left[e^{S_j}\mathbf{1}_{\{S_j \leq -r, \ \min_{1 \leq i < j} S_i > -r\}}\right] \leq e^{-r}\sum_{j=1}^{\infty} \mathbf{P}\left\{S_j \leq -r, \ \min_{1 \leq i < j} S_i > -r\right\} = e^{-r}.$$

Hence, for $r \in \mathbb{R}$,

$$\mathbf{P}\left(\inf_{y \in \mathbb{T}} V(y) \leq -r\right) \leq e^{-r}.$$
$$(5.42)$$

[The inequality holds trivially if $r \leq 0$.] Taking $r := z$, we see that

$$\mathbf{P}\left(M_n \leq \frac{3}{2}\ln n - z\right) \leq e^{-z} + \mathbf{P}\left(M_n \leq \frac{3}{2}\ln n - z, \ \inf_{y \in \mathbb{T}} V(y) > -z\right)$$

$$= e^{-z} + \mathbf{P}_z\left(M_n \leq \frac{3}{2}\ln n, \ \inf_{y \in \mathbb{T}} V(y) > 0\right).$$

The last probability expression on the right-hand side is bounded by $c_{29}\,(1 + z)e^{-z}$ (see (5.41)). Consequently, there exists a constant $c_{30} > 0$ such that for all sufficiently large n and all $z \in \mathbb{R}$ (the case $z < 0$ being trivial),

$$\mathbf{P}\left(M_n \leq \frac{3}{2}\ln n - z\right) \leq c_{30}\,(1 + z_+)\,e^{-z},$$
$$(5.43)$$

which is in agreement with (5.39). $\qquad\square$

We now have all the ingredients for the proof of Lemma 5.18.

Proof of Lemma 5.18 For the first inequality (5.31), it suffices to prove that

$$\mathbf{P}\Big(\exists x \in \mathbb{T}: \ |x| = n, \ V(x) - \frac{3}{2}\ln n \le -z, \ E_n(x)^c\Big) \le e^{K-z} + \varepsilon(1 + z - K)e^{-z},$$

for $n \ge n_0$, $L \ge L_0$ and $0 \le K \le z \le \frac{3}{2}\ln n - L$. Recall that $E_n(x) :=$ $\{\min_{0 \le i \le n} V(x_i) \ge -z + K, \ \min_{\frac{n}{2} \le j \le n} V(x_j) \ge \frac{3}{2}\ln n - z - L\}$, for $|x| = n$, as defined in (5.30).

Taking $r := z - K$ in (5.42), we see that the proof of Lemma 5.18 is reduced to showing the following:

$$p_{(5.44)} \le \varepsilon(1 + z - K)e^{-z}, \tag{5.44}$$

where

$$p_{(5.44)} := \mathbf{P}\Big(\exists x \in \mathbb{T}: \ |x| = n, \ \min_{1 \le i \le n} V(x_i) \ge -z + K,$$

$$V(x) - \frac{3}{2}\ln n \le -z, \ \min_{\frac{n}{2} \le j \le n} V(x_j) < \frac{3}{2}\ln n - z - L\Big).$$

Also, we realize that the second inequality (5.32) in the lemma is a consequence of (5.44). So the rest of the proof of the lemma is devoted to verifying (5.44). We have

$$p_{(5.44)} = \mathbf{P}_{z-K}\Big(\exists x \in \mathbb{T}: \ |x| = n, \ \min_{1 \le i \le n} V(x_i) \ge 0,$$

$$V(x) - \frac{3}{2}\ln n \le -K, \ \min_{\frac{n}{2} \le j \le n} V(x_j) < \frac{3}{2}\ln n - L - K\Big)$$

$$\le \sum_{\ell=K}^{\infty} \sum_{k=\max\{L+K, \ell\}}^{\infty} \mathbf{P}_{z-K}\Big(\exists x \in \mathbb{T}: \ |x| = n, \ \min_{1 \le i \le n} V(x_i) \ge 0,$$

$$V(x) \simeq \frac{3}{2}\ln n - \ell, \ \min_{\frac{n}{2} \le j \le n} V(x_j) \simeq \frac{3}{2}\ln n - k\Big),$$

where, as before, $a \simeq b$ is short for $|a - b| \le 1$. By Lemma 5.20, this yields

$$p_{(5.44)} \le \sum_{\ell=K}^{\infty} \sum_{k=\max\{L+K, \ell\}}^{\infty} c_{20} (1 + z - K)e^{-c_{21}(k-\ell)-(z-K)-\ell}.$$

Without loss of generality, we can assume $c_{21} < 1$; otherwise, we replace it by $\min\{c_{21}, \frac{1}{2}\}$. This gives, with a change of indices $\ell' := \ell - K$ and $k' := k - K$, that

$$p_{(5.44)} \le c_{20} (1 + z - K) e^{-z} \sum_{k'=L}^{\infty} \sum_{\ell'=0}^{k'} e^{-c_{21}(k'-\ell')-\ell'}.$$

But, $\sum_{k'=L}^{\infty}\sum_{\ell'=0}^{k'} e^{-c_{21}(k'-\ell')-\ell'} \leq \sum_{k'=L}^{\infty} e^{-c_{21}k'} \sum_{\ell'=0}^{\infty} e^{-(1-c_{21})\ell'}$, which we write as $c_{31}\sum_{k'=L}^{\infty} e^{-c_{21}k'}$, with $c_{31} := \sum_{\ell'=0}^{\infty} e^{-(1-c_{21})\ell'} < \infty$. Since $\sum_{k'=L}^{\infty} e^{-c_{21}k'}$ can be made as small as possible for all $L \geq L_0$ as long as L_0 is chosen to be sufficiently large, this completes the proof of Lemma 5.18.   □

5.4.4 Step 3b: Proof of Lemma 5.19

Let $\mathscr{L}_n^{z,L,K}$ and $\mathscr{Y}_n^{z,L,K}$ be as in (5.17) and (5.20), respectively. Fix $\eta > 0$ and $L \geq 0$. By the peeling lemma (Theorem 5.14 in Sect. 5.3), for all sufficiently large n, and all $0 \leq K \leq z \leq \frac{3}{2}\ln n - L$,

$$\mathbf{Q}\left(w_n \in \mathscr{L}_n^{z,L,K} \backslash \mathscr{Y}_n^{z,L,K}\right) \leq \frac{\eta(1+z-K)}{2\,n^{3/2}}.$$

Since $E_n(w_n) \subset \{w_n \in \mathscr{L}_n^{z,L,K}\}$ (the two sets are almost identical, the only difference being that on $E_n(w_n)$, we have $V(w_n) \leq \frac{3}{2}\ln n - z$, whereas on $\{w_n \in \mathscr{L}_n^{z,L,K}\}$, we have $V(w_n) \leq \frac{3}{2}\ln n - z + C$), it remains to prove that

$$\mathbf{Q}\left(E_n(w_n),\ \mathscr{E}_{n,b}^c,\ w_n \in \mathscr{Y}_n^{z,L,K}\right) \leq \frac{\eta(1+z-K)}{2\,n^{3/2}}. \tag{5.45}$$

We make the following simple observation: On the event $\{w_n \in \mathscr{Y}_n^{z,L,K}\}$ (for the definition of $\mathscr{Y}_n^{z,L,K}$, see (5.20)), for any $k \leq n$ and $y \in \mathrm{brot}(w_k)$, we have $e^{-[V(y)-a_k^{(n)}]} \leq \rho\,e^{-\beta_k^{(n)}} \leq \rho$, so

$$V(y) \geq a_k^{(n)} - \ln \rho. \tag{5.46}$$

Let $\mathscr{G}_n := \sigma\{w_i,\ V(w_i),\ \mathrm{brot}(w_i),\ (V(y))_{y\in\mathrm{brot}(w_i)},\ 1 \leq i \leq n\}$ be the σ-field generated by the spine and its children in the first n generations. Clearly, $E_n(w_n)$ and $\{w_n \in \mathscr{Y}_n^{z,L,K}\}$ are both $\mathscr{G}_n$-measurable. We have

$$\mathbf{Q}\left(\mathscr{E}_{n,b}^c \mid \mathscr{G}_n\right) = 1 - \prod_{k=1}^{n-b}\ \prod_{y\in\mathrm{brot}(w_k)} [1 - \Psi_{k,n,z}(V(y))], \tag{5.47}$$

where

$$\Psi_{k,n,z}(r) := \mathbf{P}\left(M_{n-k} \leq \frac{3}{2}\ln n - z - r\right).$$

We first look at the situation $k \leq \frac{n}{2}$ (so $a_k^{(n)} = -z + K$). Since $\ln n \leq \ln(n-k) + \ln 2$, it follows from (5.43) that

$$\Psi_{k,n,z}(r) \leq c_{32} \left(1 + (r+z)_+\right) e^{-r-z} \leq c_{33} \, K e^{-K} \left[1 + (r - a_k^{(n)})_+\right] e^{-(r-a_k^{(n)})}.$$

Let $y \in \mathrm{brot}(w_k)$. We have noted in (5.46) that on $\{w_n \in \mathscr{Y}_n^{z,L,K}\}$, $V(y) - a_k^{(n)} \geq -\ln \rho$, so it is possible to choose and fix K_1 sufficiently large such that for $K \geq K_1$, $c_{33} \, K e^{-K} \left[1 + (V(y) - a_k^{(n)})_+\right] e^{-[V(y) - a_k^{(n)}]} \leq \frac{1}{2}$. By the elementary inequality $1 - u \geq e^{-c_{34} u}$ for some $c_{34} > 0$ and all $u \in [0, \frac{1}{2}]$, we see that for $k \leq \frac{n}{2}$,

$$\prod_{y \in \mathrm{brot}(w_k)} [1 - \Psi_{k,n,z}(V(y))]$$

$$\geq \prod_{y \in \mathrm{brot}(w_k)} \{1 - c_{33} \, K e^{-K} \left[1 + (V(y) - a_k^{(n)})_+\right] e^{-[V(y) - a_k^{(n)}]}\},$$

$$\geq \exp\left(-c_{34} c_{33} \, K e^{-K} \sum_{y \in \mathrm{brot}(w_k)} \left[1 + (V(y) - a_k^{(n)})_+\right] e^{-[V(y) - a_k^{(n)}]}\right).$$

On $\{w_n \in \mathscr{Y}_n^{z,L,K}\}$, we have, for $k \leq \frac{n}{2}$,

$$\sum_{y \in \mathrm{brot}(w_k)} \left[1 + (V(y) - a_k^{(n)})_+\right] e^{-[V(y) - a_k^{(n)}]} \leq \rho e^{-\beta_k^{(n)}} = \rho e^{-k^{1/7}},$$

which yields (writing $c_{35} := c_{34} c_{33} \rho$), on $\{w_n \in \mathscr{Y}_n^{z,L,K}\}$,

$$\prod_{k=1}^{\frac{n}{2}} \prod_{y \in \mathrm{brot}(w_k)} [1 - \Psi_{k,n,z}(V(y))] \geq \exp\left(-c_{35} \, K e^{-K} \sum_{k=1}^{\frac{n}{2}} e^{-k^{1/7}}\right)$$

$$\geq \exp\left(-c_{36} \, K e^{-K}\right)$$

$$\geq (1 - \eta)^{1/2}, \tag{5.48}$$

for all sufficiently large K ($\eta \in (0, 1)$ being fixed), where $c_{36} := c_{35} \sum_{k=1}^{\infty} e^{-k^{1/7}}$.

We still need to take care of the product $\prod_{k=\frac{n}{2}+1}^{n-b}$. Recall that $\Psi_{k,n,z}(r) = \mathbf{P}(M_{n-k} \leq \frac{3}{2} \ln n - z - r)$. Let $\frac{n}{2} < k \leq n - b$ (so $a_k^{(n)} = \frac{3}{2} \ln n - z - L$), and let $y \in \mathrm{brot}(w_k)$. On $\{w_n \in \mathscr{Y}_n^{z,L,K}\}$, we have observed in (5.46) that $V(y) \geq a_k^{(n)} - \ln \rho = \frac{3}{2} \ln n - z - L - \ln \rho$, i.e., $\frac{3}{2} \ln n - z - V(y) \leq L + \ln \rho$ which is a given constant (recalling that $L \geq 0$ is fixed). Since $n - k \geq b$, this yields that $\Psi_{k,n,z}(V(y))$ can be as small as possible (on $\{w_n \in \mathscr{Y}_n^{z,L,K}\}$) if b is chosen to be large; in particular, $\Psi_{k,n,z}(V(y)) \leq \frac{1}{2}$, so again

$$\prod_{k=\frac{n}{2}+1}^{n-b} \prod_{y \in \mathrm{brot}(w_k)} [1 - \Psi_{k,n,z}(V(y))] \geq \exp\left(-c_{34} \sum_{k=\frac{n}{2}+1}^{n-b} \sum_{y \in \mathrm{brot}(w_k)} \Psi_{k,n,z}(V(y))\right).$$

By (5.42),

$$\Psi_{k,n,z}(r) \leq \mathbf{P}\left(\inf_{x \in \mathbb{T}} V(x) \leq \frac{3}{2}\ln n - z - r\right) \leq e^{\frac{3}{2}\ln n - z - r}.$$

Consequently, on $\{w_n \in \mathscr{Y}_n^{z,L,K}\}$, for $\frac{n}{2} < k \leq n - b$,

$$\sum_{y \in \mathrm{brot}(w_k)} \Psi_{k,n,z}(V(y)) \leq \sum_{y \in \mathrm{brot}(w_k)} e^{\frac{3}{2}\ln n - z - V(y)}$$

$$= e^L \sum_{y \in \mathrm{brot}(w_k)} e^{-(V(y) - a_k^{(n)})}$$

$$\leq e^L \rho\, e^{-(n-k)^{1/7}},$$

which yields that, on $\{w_n \in \mathscr{Y}_n^{z,L,K}\}$,

$$\prod_{k=\frac{n}{2}+1}^{n-b} \prod_{y \in \mathrm{brot}(w_k)} [1 - \Psi_{k,n,z}(V(y))] \geq \exp\left(-c_{34}e^L \rho \sum_{k=\frac{n}{2}+1}^{n-b} e^{-(n-k)^{1/7}}\right),$$

which is greater than $(1 - \eta)^{1/2}$ if b is sufficiently large (recalling that $L \geq 0$ and $\eta \in (0, 1)$ are fixed). Together with (5.48), we see that for K and b sufficiently large,

$$\prod_{k=1}^{n-b} \prod_{y \in \mathrm{brot}(w_k)} [1 - \Psi_{k,n,z}(V(y))] \geq 1 - \eta.$$

Going back to (5.47) yields that for K and b sufficiently large,

$$\mathbf{Q}\left(E_n(w_n),\, \mathscr{E}_{n,b}^c,\, w_n \in \mathscr{Y}_n^{z,L,K}\right) \leq \eta\, \mathbf{Q}\left(E_n(w_n),\, w_n \in \mathscr{Y}_n^{z,L,K}\right) \leq \eta\, \mathbf{Q}\left(E_n(w_n)\right).$$

Since $\mathbf{Q}(E_n(w_n)) = \mathbf{P}(\min_{0 \leq i \leq n} S_i \geq -z + K,\ \min_{\frac{n}{2} \leq j \leq n} S_j \geq \frac{3}{2}\ln n - z - L,\ S_n \leq \frac{3}{2}\ln n - z)$, which is bounded by $c_{68}\,(L+1)^2 \frac{1+z-K}{n^{3/2}}$ by Lemma A.4 of Appendix A.2 (where c_{68} is the constant in Lemma A.4). Consequently,

$$\mathbf{Q}\left(E_n(w_n),\, \mathscr{E}_{n,b}^c,\, w_n \in \mathscr{Y}_n^{z,L,K}\right) \leq \eta\, c_{68}\,(L+1)^2 \frac{1+z-K}{n^{3/2}},$$

for fixed $0 < \eta < 1$, $L \geq 0$ and all sufficiently large n, $b < n$ and K (satisfying $K \leq z \leq \frac{3}{2} \ln n - L$). Since η can be arbitrarily small, we should have worked with $\frac{\eta}{2c_{68}(L+1)^2}$ in place of η to obtain (5.45). Lemma 5.19 is proved. □

5.4.5 Step 4: The Role of the Non-lattice Assumption

We have already stated that Theorem 5.15 fails without the assumption that the law of Ξ is non-lattice. Concretely, this assumption is needed in Proposition 5.22 below, which is the last technical estimate in our proof of Theorem 5.15.

Loosely speaking, the branching random walk with absorption along the path $[\![\varnothing, m^{(n)}]\!]$, behaves like a centred random walk conditioned to stay above 0 during the first $\frac{n}{2}$ steps and above $\frac{3}{2} \ln n + O(1)$ during the last $\frac{n}{2}$ steps. Proposition 5.22 below describes the distribution of such a random walk. Throughout this subsection, let (S_j) denote a centred random walk with $\sigma^2 := \mathbf{E}(S_1^2) \in (0, \infty)$.

Let R be the renewal function defined in the sense of (A.1) (see Appendix A.1). Let R_- be the renewal function associated with the random walk $(-S_n)$. Write

$$\underline{S}_n := \min_{0 \leq i \leq n} S_i, \quad n \geq 0.$$

Proposition 5.22 (Random Walk Above a Barrier) *Let (r_n) be a sequence of positive real numbers such that $\frac{r_n}{n^{1/2}} \to 0$, $n \to \infty$. Let (λ_n) be such that $0 < \liminf_{n \to \infty} \lambda_n < \limsup_{n \to \infty} \lambda_n < 1$. Let $a \geq 0$. Let $F : \mathbb{R}_+ \to \mathbb{R}_+$ be a Riemann-integrable function such that there exists a non-increasing function $F_1 : [0, \infty) \to \mathbb{R}$ satisfying $|F| \leq F_1$ and $\int_0^\infty u F_1(u) \, du < \infty$. If the distribution of S_1 is non-lattice, then*

$$\lim_{n \to \infty} n^{3/2} \, \mathbf{E}\left[F(S_n - y) \, \mathbf{1}_{\{\underline{S}_n \geq -a\}} \, \mathbf{1}_{\{\min_{\lambda_n n \leq i \leq n} S_i \geq y\}} \right]$$

$$= C_+ C_- (\frac{\pi}{2\sigma^2})^{1/2} R(a) \int_0^\infty F(u) R_-(u) \, du, \tag{5.49}$$

uniformly in $y \in [0, r_n]$, where C_+ and C_- are the constants in (A.7) of Appendix A.2.

Proof By considering the positive and negative parts, we clearly can assume that $F \geq 0$. Since the limit on the right-hand side of (5.49) does not depend on (λ_n), we can also assume without loss of generality (using monotonicity) that $\lambda_n = \lambda$ for all n and some $\lambda \in (0, 1)$.

We now argue that only functions F with compact support need to be taken care of. Let $\varepsilon > 0$. Let $M \geq 1$ be an integer. By assumption, $F \leq F_1$ and F_1 is non-increasing; so (notation: $u \simeq v$ is again short for $|u - v| \le 1$)

$$\mathbf{E}\Big[F(S_n - y)\,\mathbf{1}_{\{S_n - y > M\}}\,\mathbf{1}_{\{\underline{S}_n \ge -a\}}\,\mathbf{1}_{\{\min_{\lambda n \le i \le n} S_i \ge y\}}\Big]$$

$$\le \sum_{k=M}^{\infty} F_1(k-1)\,\mathbf{P}\Big(S_n \simeq y + k,\ \underline{S}_n \ge -a,\ \min_{\lambda n \le i \le n} S_i \ge y\Big),$$

which is bounded by $\sum_{k=M}^{\infty} F_1(k-1)\,c_{37}\,\frac{(k+2)(a+1)}{n^{3/2}}$ (Lemma A.4 of Appendix A.2). By assumption, $\int_{\mathbb{R}_+} x\,F_1(x)\,\mathrm{d}x < \infty$, so $\sum_{k=M}^{\infty}(k+2)\,F_1(k-1) \to 0$ if $M \to \infty$; in particular, for any $\varepsilon > 0$, we can choose the integer $M \ge 1$ sufficiently large such that $n^{3/2}\,\mathbf{E}[F(S_n - y)\,\mathbf{1}_{\{S_n - y > M\}}\,\mathbf{1}_{\{\underline{S}_n \ge -a\}}\,\mathbf{1}_{\{\min_{\lambda n \le i \le n} S_i \ge y\}}] < \varepsilon$.

So we only need to treat function F with compact support. By assumption, F is Riemann-integrable, so by approximating F with step functions, we only need to prove the proposition for $F(u) := \mathbf{1}_{[0,\chi]}(u)$ (where $\chi > 0$ is a fixed constant). For such F, denoting by $\mathbf{E}_{(5.49)}$ the expectation on the left-hand side of (5.49),

$$\mathbf{E}_{(5.49)} = \mathbf{P}\Big(\underline{S}_n \ge -a,\ \min_{\lambda n \le i \le n} S_i \ge y,\ S_n \le y + \chi\Big).$$

Applying the Markov property at time λn gives

$$\mathbf{E}_{(5.49)} = \mathbf{E}\Big[f_{(5.50)}(S_{\lambda n})\,\mathbf{1}_{\{\underline{S}_{\lambda n} \ge -a\}}\Big],$$

where, for $u \ge 0$,

$$f_{(5.50)}(u) = f_{(5.50)}(u, n, y, \chi) := \mathbf{P}_u\Big(\underline{S}_{(1-\lambda)n} \ge y,\ S_{(1-\lambda)n} \le y + \chi\Big). \tag{5.50}$$

For notational simplification, we write $n_1 := (1 - \lambda)n$. Since $(S_{n_1} - S_{n_1 - i},\ 0 \le i \le n_1)$ is distributed as $(S_i,\ 0 \le i \le n_1)$, we have $f_{(5.50)}(u) = \mathbf{P}\{\underline{S}_{n_1} \ge (-S_{n_1}) + (y - u) \ge -\chi\}$, where $\underline{S}_j^- := \min_{0 \le i \le j}(-S_i)$.

Let $\tau_{n_1} := \min\{j:\ 0 \le j \le n_1,\ -S_j = \underline{S}_{n_1}^-\}$. In words, τ_{n_1} is the first time $(-S_i)$ hits its minimum during $[0, n_1]$. We have

$$f_{(5.50)}(u) = \sum_{j=0}^{n_1} \mathbf{P}\Big(\tau_{n_1} = j,\ \underline{S}_{n_1}^- \ge (-S_{n_1}) + (y - u) \ge -\chi\Big).$$

Applying the Markov property at time j yields that

$$f_{(5.50)}(u) = \sum_{j=0}^{n_1} \mathbf{E}\Big[g_{(5.52)}(u - y,\ \underline{S}_j^- + \chi,\ n_1 - j)\,\mathbf{1}_{\{-S_j = \underline{S}_j^- \ge -\chi\}}\Big], \tag{5.51}$$

where, for $z \ge 0$, $v \ge 0$ and $\ell \ge 0$,

$$g_{(5.52)}(z, v, \ell) := \mathbf{P}\Big(z - v \le -S_\ell \le z,\ \underline{S}_\ell^- \ge 0\Big). \tag{5.52}$$

Write $\psi(u) := u\,e^{-u^2/2}\,\mathbf{1}_{\{u\geq 0\}}$. Recall our notation $\sigma^2 := \mathbf{E}(S_1^2)$. By Theorem 1 of Caravenna [82] (this is where the non-lattice assumption is needed; the lattice case is also studied in [82], see Theorem 2 there), for $\ell \to \infty$,

$$\mathbf{P}\!\left(z - v \leq -S_\ell \leq z \,\middle|\, \underline{S_\ell} \geq 0\right) = \frac{v}{\sigma\ell^{1/2}}\,\psi\!\left(\frac{z}{\sigma\ell^{1/2}}\right) + o\!\left(\frac{1}{\ell^{1/2}}\right),$$

uniformly in $z \geq 0$ and in v in any compact subset of $[0, \infty)$. Multiplying both sides by $\mathbf{P}(\underline{S_\ell} \geq 0)$, which is equivalent to $\frac{C_-}{\ell^{1/2}}$ (see (A.7) in Appendix A.2), we get, for $\ell \to \infty$,

$$g_{(5.52)}(z, v, \ell) = C_- \frac{v}{\sigma\ell}\,\psi\!\left(\frac{z}{\sigma\ell^{1/2}}\right) + o\!\left(\frac{1}{\ell}\right), \tag{5.53}$$

uniformly in $z \geq 0$ and in $v \in [0, \chi]$. In particular, since ψ is bounded, there exists a constant $c_{38} = c_{38}(\chi) > 0$ such that for all $z \geq 0$, $\ell \geq 0$ and $v \in [0, \chi]$,

$$g_{(5.52)}(z, v, \ell) \leq \frac{c_{38}}{\ell + 1}.$$

We now return to (5.51) and continue our study of $f_{(5.50)}(u)$. We split the sum $\sum_{j=0}^{n_1}$ into $\sum_{j=0}^{j_n} + \sum_{j=j_n+1}^{n_1}$, where $j_n := \lfloor n^{1/2} \rfloor$. Since $n_1 := (1-\lambda)n$ by definition, we have $j_n < n_1$ for all sufficiently large n. As such,

$$f_{(5.50)}(u) = f_{(5.50)}^{(1)}(u) + f_{(5.50)}^{(2)}(u),$$

where

$$f_{(5.50)}^{(1)}(u) := \sum_{j=0}^{j_n} \mathbf{E}\!\left[g_{(5.52)}(u - y,\, \underline{S_j^-} + \chi,\, n_1 - j)\,\mathbf{1}_{\{-S_j = \underline{S_j^-} \geq -\chi\}}\right],$$

$$f_{(5.50)}^{(2)}(u) := \sum_{j=j_n+1}^{n_1} \mathbf{E}\!\left[g_{(5.52)}(u - y,\, \underline{S_j^-} + \chi,\, n_1 - j)\,\mathbf{1}_{\{-S_j = \underline{S_j^-} \geq -\chi\}}\right].$$

We let $n \to \infty$. Note that ψ is uniformly continuous and bounded on $[0, \infty)$. By (5.53), as long as $y = o(n^{1/2})$,

$$f_{(5.50)}^{(1)}(u) = \frac{C_-}{\sigma n_1}\,(1 + o(1))\,\psi\!\left(\frac{u}{\sigma n_1^{1/2}}\right) \sum_{j=0}^{j_n} \mathbf{E}\!\left[(\underline{S_j^-} + \chi)\,\mathbf{1}_{\{-S_j = \underline{S_j^-} \geq -\chi\}}\right]$$

$$+ o\!\left(\frac{1}{n}\right) \sum_{j=0}^{j_n} \mathbf{P}\!\left(-S_j = \underline{S_j^-} \geq -\chi\right).$$

Since $\sum_{j=0}^{\infty} \mathbf{E}[(\underline{S}_j^- + \chi)\, \mathbf{1}_{\{-S_j = \underline{S}_j^- \geq -\chi\}}] = \int_0^\chi R_-(t)\,dt$ and $\sum_{j=0}^{\infty} \mathbf{P}(-S_j = \underline{S}_j^- \geq -\chi) = R_-(\chi)$, this yields, for $y = o(n^{1/2})$,

$$f_{(5.50)}^{(1)}(u) = \frac{C_-}{\sigma n_1}\, \psi\left(\frac{u}{\sigma n_1^{1/2}}\right) \int_0^\chi R_-(t)\,dt + o\left(\frac{1}{n}\right). \qquad (5.54)$$

To treat $f_{(5.50)}^{(2)}(u)$, we simply use the inequality $g_{(5.52)}(z, v, \ell) \leq \frac{c_{38}}{\ell+1}$ (for $z \geq 0$, $\ell \geq 0$ and $v \in [0, \chi]$), to see that

$$f_{(5.50)}^{(2)}(u) \leq c_{38} \sum_{j=j_n+1}^{n_1} \frac{1}{n_1 - j + 1} \mathbf{P}\left(\underline{S}_j^- \geq -\chi,\ -S_j \leq 0\right),$$

which, by Lemma A.1 of Appendix A.2, is bounded by $c_{39} \sum_{j=j_n+1}^{n_1} \frac{1}{(n_1-j+1) j^{3/2}}$, which is $o(\frac{1}{n})$. Together with (5.54), we get, as $n \to \infty$,

$$f_{(5.50)}(u) = \frac{C_-}{\sigma(1-\lambda)n}\, \psi\left(\frac{u}{\sigma[(1-\lambda)n]^{1/2}}\right) \int_0^\chi R_-(t)\,dt + o\left(\frac{1}{n}\right),$$

uniformly in $u \geq 0$ and $y \in [0, r_n]$. Since $\mathbf{E}_{(5.49)} = \mathbf{E}[f_{(5.50)}(S_{\lambda n})\,\mathbf{1}_{\{S_{\lambda n} \geq -a\}}]$, and $\mathbf{P}(\underline{S}_\ell \geq -a) \sim \frac{C_+ R(a)}{\ell^{1/2}}$, $\ell \to \infty$ (see (A.7) in Appendix A.2), this yields

$$\mathbf{E}_{(5.49)} = \frac{C_-}{\sigma(1-\lambda)n} \int_0^\chi R_-(t)\,dt\, \frac{C_+ R(a)}{(\lambda n)^{1/2}}\, \mathbf{E}_a\left[\psi\left(\frac{S_{\lambda n} - a}{\sigma[(1-\lambda)n]^{1/2}}\right)\,\Big|\, \underline{S}_{\lambda n} \geq 0\right]$$
$$+ o\left(\frac{1}{n^{3/2}}\right).$$

Under the conditional probability $\mathbf{P}_a(\cdot \mid \underline{S}_{\lambda n} \geq 0)$, $\frac{S_\ell}{\sigma\, \ell^{1/2}}$ converges weakly (as $\ell \to \infty$) to the Rayleigh distribution, whose density is ψ (see [83]). Hence

$$\lim_{n\to\infty} \mathbf{E}_a\left[\psi\left(\frac{S_{\lambda n} - a}{\sigma[(1-\lambda)n]^{1/2}}\right)\,\Big|\, \underline{S}_{\lambda n} \geq 0\right] = \int_0^\infty \psi\left(\frac{\lambda^{1/2}}{(1-\lambda)^{1/2}}\, t\right) \psi(t)\,dt,$$

which is equal to $\lambda^{1/2}(1-\lambda)(\frac{\pi}{2})^{1/2}$. Since $\int_0^\chi R_-(t)\,dt = \int_0^\infty R_-(t)F(t)\,dt$ with $F := \mathbf{1}_{[0,\,\chi]}$, this yields Proposition 5.22. $\square$

5.5 Leftmost Position: Fluctuations

Consider a branching random walk under Assumption (H). Theorem 5.12 in Sect. 5.3 says that under $\mathbf{P}^*(\cdot) := \mathbf{P}(\cdot \mid \text{non-extinction})$,

$$\frac{1}{\ln n}\, \inf_{|x|=n} V(x) \to \frac{3}{2}, \quad \text{in probability.}$$

[Of course, Theorem 5.15 in Sect. 5.4 tells us that a lot more is true if the law of the underlying point process is non-lattice.] It is nice that the minimal position has such a strong universality. The aim of this section, however, is to show that we cannot go further. In fact, we are going to see that $\mathbf{P}^*$-almost surely, for any[8] $\varepsilon > 0$, there exists an exceptional subsequence along which $\inf_{|x|=n} V(x)$ goes below $(\frac{1}{2}+\varepsilon) \ln n$.

Theorem 5.23 *Under Assumption* (H), *we have, under* $\mathbf{P}^*$,

$$\lim_{n\to\infty} \frac{1}{\ln n} \inf_{|x|=n} V(x) = \frac{3}{2}, \quad \text{in probability,} \tag{5.55}$$

$$\limsup_{n\to\infty} \frac{1}{\ln n} \inf_{|x|=n} V(x) = \frac{3}{2}, \quad \text{a.s.} \tag{5.56}$$

$$\liminf_{n\to\infty} \frac{1}{\ln n} \inf_{|x|=n} V(x) = \frac{1}{2}, \quad \text{a.s.} \tag{5.57}$$

Of course, in Theorem 5.23, the convergence in probability, (5.55), is just a restatement of Theorem 5.12 made for the sake of completeness. Only (5.56) and (5.57) are new.

The lower bound in (5.57) can be strengthened as follows.

Theorem 5.24 *Under Assumption* (H), *we have*

$$\liminf_{n\to\infty} \left(\inf_{|x|=n} V(x) - \frac{1}{2} \ln n \right) = -\infty, \quad \mathbf{P}^*\text{-a.s.}$$

The proof of Theorem 5.24 relies on the following estimate. Let $C > 0$ be the constant in Lemma A.10 (Appendix A.2).

Lemma 5.25 *Under Assumption* (H),

$$\liminf_{n\to\infty} \mathbf{P}\left\{ \exists x : n \le |x| \le 2n, \ \frac{1}{2} \ln n \le V(x) \le \frac{1}{2} \ln n + C \right\} > 0.$$

Proof of Lemma 5.25 The proof is similar to the proof of Lemma 5.13 presented in Sect. 5.3 (in the special case $z = 0$, which simplifies the writing), by a second moment method and using the peeling lemma (Theorem 5.14 in Sect. 5.3), but this time we count all the generations k between n et $2n$ (instead of just generation $k = n$ in Lemma 5.13). This explains the factor $\frac{1}{2}$ instead of $\frac{3}{2}$, because $\sum_{k=n}^{2n} \frac{1}{n^{3/2}} \sim \frac{1}{n^{1/2}}$, where $\frac{1}{n^{3/2}}$ comes from the probability estimate in (5.18) or (5.21). We feel free to omit the details, and refer the interested reader to [15]. □

Proof of Theorem 5.24 Let $\chi > 0$. The system being supercritical, the assumption $\psi'(1) = 0$ ensures $\mathbf{P}\{\inf_{|x|=1} V(x) < 0\} > 0$. Therefore, there exists an integer

---

[8] Actually, it also holds if $\varepsilon = 0$. See Theorem 5.24 below.

$L = L(\chi) \geq 1$ such that

$$c_{40} := \mathbf{P}\left\{ \inf_{|x|=L} V(x) \leq -\chi \right\} > 0.$$

Let $n_k := (L+2)^k$, $k \geq 1$, so that $n_{k+1} \geq 2n_k + L$, $\forall k$. For any k, let

$$T_k := \inf\left\{ i \geq n_k : \inf_{|x|=i} V(x) \leq \frac{1}{2} \ln n_k + C \right\},$$

where $C > 0$ is the constant in Lemma 5.25. If $T_k < \infty$, let x_k be such that[9] $|x_k| = T_k$ and that $V(x) \leq \frac{1}{2} \ln n_k + C$. Let

$$\mathscr{A}_k := \{T_k \leq 2n_k\} \cap \left\{ \inf_{y>x_k:\, |y|=|x_k|+L} [V(y) - V(x_k)] \leq -\chi \right\},$$

where $y > x_k$ means, as before, that y is a descendant of x_k. For any pair of positive integers $j < \ell$,

$$\mathbf{P}\left\{ \bigcup_{k=j}^{\ell} \mathscr{A}_k \right\} = \mathbf{P}\left\{ \bigcup_{k=j}^{\ell-1} \mathscr{A}_k \right\} + \mathbf{P}\left\{ \bigcap_{k=j}^{\ell-1} \mathscr{A}_k^c \cap \mathscr{A}_\ell \right\}. \tag{5.58}$$

On $\{T_\ell < \infty\}$, we have

$$\mathbf{P}\{\mathscr{A}_\ell \mid \mathscr{F}_{T_\ell}\} = \mathbf{1}_{\{T_\ell \leq 2n_\ell\}}\, \mathbf{P}\left\{ \inf_{|x|=L} V(x) \leq -\chi \right\} = c_{41}\, \mathbf{1}_{\{T_\ell \leq 2n_\ell\}}.$$

Since $\cap_{k=j}^{\ell-1} \mathscr{A}_k^c$ is $\mathscr{F}_{T_\ell}$-measurable, we obtain:

$$\mathbf{P}\left\{ \bigcap_{k=j}^{\ell-1} \mathscr{A}_k^c \cap \mathscr{A}_\ell \right\} = c_{41}\, \mathbf{P}\left\{ \bigcap_{k=j}^{\ell-1} \mathscr{A}_k^c \cap \{T_\ell \leq 2n_\ell\} \right\}$$

$$\geq c_{41}\, \mathbf{P}\{T_\ell \leq 2n_\ell\} - c_{41}\, \mathbf{P}\left\{ \bigcup_{k=j}^{\ell-1} \mathscr{A}_k \right\}.$$

Recall that $\mathbf{P}\{T_\ell \leq 2n_\ell\} \geq c_{42}$ (Lemma 5.25; for large ℓ, say $\ell \geq j_0$). Combining this with (5.58) yields that

$$\mathbf{P}\left\{ \bigcup_{k=j}^{\ell} \mathscr{A}_k \right\} \geq (1 - c_{41})\mathbf{P}\left\{ \bigcup_{k=j}^{\ell-1} \mathscr{A}_k \right\} + c_{42}c_{41}, \quad j_0 \leq j < \ell.$$

---

[9]If the choice of x_k is not unique, we can choose for example the one with the smallest Harris–Ulam index.

Iterating the inequality leads to:

$$\mathbf{P}\left\{\bigcup_{k=j}^{\ell} \mathscr{A}_k\right\} \geq (1 - c_{41})^{\ell-j}\, \mathbf{P}\{\mathscr{A}_j\} + c_{42}c_{41} \sum_{i=0}^{\ell-j-1}(1 - c_{41})^i \geq c_{42}c_{41} \sum_{i=0}^{\ell-j-1}(1 - c_{41})^i.$$

This yields $\mathbf{P}\{\bigcup_{k=j}^{\infty} \mathscr{A}_k\} \geq c_{42}$, $\forall j \geq j_0$. Consequently, $\mathbf{P}(\limsup_{k\to\infty} \mathscr{A}_k) \geq c_{42}$.

On the event $\limsup_{k\to\infty} \mathscr{A}_k$, there are infinitely many vertices x such that $V(x) \leq \frac{1}{2} \ln |x| + C - \chi$. Therefore,

$$\mathbf{P}\left\{\liminf_{n\to\infty}\left(\inf_{|x|=n} V(x) - \frac{1}{2}\ln n\right) \leq C - \chi\right\} \geq c_{42}.$$

The constant $\chi > 0$ being arbitrary, we obtain:

$$\mathbf{P}\left\{\liminf_{n\to\infty}\left(\inf_{|x|=n} V(x) - \frac{1}{2}\ln n\right) = -\infty\right\} \geq c_{42}.$$

Let $0 < \varepsilon < 1$. Let $J_1 \geq 1$ be an integer such that $(1 - c_{42})^{J_1} \leq \varepsilon$. Under $\mathbf{P}^*$, the system survives almost surely; so there exists a positive integer J_2 sufficiently large such that $\mathbf{P}^*\{\sum_{|x|=J_2} 1 \geq J_1\} \geq 1 - \varepsilon$. By applying what we have just proved to the subtrees of the vertices at generation J_2, we obtain:

$$\mathbf{P}^*\left\{\liminf_{n\to\infty}\left(\inf_{|x|=n} V(x) - \frac{1}{2}\ln n\right) = -\infty\right\} \geq 1 - (1 - c_{42})^{J_1} - \varepsilon \geq 1 - 2\varepsilon.$$

Sending ε to 0 completes the proof of Theorem 5.24. $\qquad\square$

Proof of Theorem 5.23 We first check (5.57). Its upper bound being a straightforward consequence of Theorem 5.24, we only need to check the lower bound, namely, $\liminf_{n\to\infty} \frac{1}{\ln n} \inf_{|x|=n} V(x) \geq \frac{1}{2}$, $\mathbf{P}^*$-a.s.

Fix any $k > 0$ and $a < \frac{1}{2}$. By the many-to-one formula (Theorem 1.1 in Sect. 1.3),

$$\mathbf{E}\left(\sum_{|x|=n} \mathbf{1}_{\{\underline{V}(x)>-k\}} \mathbf{1}_{\{V(x)\leq a\ln n\}}\right) = \mathbf{E}\left(e^{S_n} \mathbf{1}_{\{\underline{S}_n>-k\}} \mathbf{1}_{\{S_n\leq a\ln n\}}\right)$$

$$\leq n^a\, \mathbf{P}\left(\underline{S}_n > -k,\ S_n \leq a\ln n\right),$$

which, by Lemma A.1 of Appendix A.2, is bounded by a constant multiple of $n^a \frac{(\ln n)^2}{n^{3/2}}$, and which is summable in n if $a < \frac{1}{2}$. Therefore, as long as $a < \frac{1}{2}$, we have

$$\sum_{n\geq 1}\sum_{|x|=n} \mathbf{1}_{\{\underline{V}(x)>-k\}} \mathbf{1}_{\{V(x)\leq a\ln n\}} < \infty, \qquad \mathbf{P}\text{-a.s.}$$

By Lemma 3.1 of Sect. 3.1, $\inf_{|x|=n} V(x) \to \infty$ $\mathbf{P}^*$-a.s.; thus $\inf_{|x|\geq 0} V(x) > -\infty$
$\mathbf{P}^*$-a.s. Consequently, $\lim\inf_{n\to\infty} \frac{1}{\ln n} \inf_{|x|=n} V(x) \geq a$, $\mathbf{P}^*$-a.s., for any $a < \frac{1}{2}$. This
yields the desired lower bound in (5.57).

It remains to prove (5.56). Its lower bound follows immediately from
Theorem 5.12 (convergence in probability implying almost sure consequence along a subsequence), so we only need to show the upper bound:
$\lim\sup_{n\to\infty} \frac{1}{\ln n} \inf_{|x|=n} V(x) \leq \frac{3}{2}$, $\mathbf{P}^*$-a.s. This, however, is not the best known
result. A much more precise result can be stated, see Theorem 5.26 below under a
slightly stronger condition.   □

Theorem 5.26 (Hu [132]) *Under Assumption* (H), *if* $\mathbf{E}[\sum_{|x|=1}(V(x)_+)^3 e^{-V(x)}] <$
∞,[10] *then*

$$\limsup_{n\to\infty} \frac{1}{\ln\ln\ln n}\left(\inf_{|x|=n} V(x) - \frac{3}{2}\ln n\right) = 1 \quad \mathbf{P}^*\text{-a.s.}$$

Theorem 5.23 says that $\lim\inf_{n\to\infty} [\inf_{|x|=n} V(x) - \frac{1}{2}\ln n] = -\infty$, $\mathbf{P}^*$-a.s., but
its proof fails to give information about how this "lim inf" expression goes to $-\infty$.
Here is a natural question.

Question 5.27 Is there a deterministic sequence (a_n) with $\lim_{n\to\infty} a_n = \infty$ such
that

$$-\infty < \liminf_{n\to\infty} \frac{1}{a_n}\left(\inf_{|x|=n} V(x) - \frac{1}{2}\ln n\right) < 0, \quad \mathbf{P}^*\text{-a.s.?}$$

I suspect that the answer to Question 5.27 is "no". If so, it would be interesting
to answer the following question.

Question 5.28 Let (a_n) be a non-decreasing sequence such that $\lim_{n\to\infty} a_n = \infty$.
Give an integral criterion on (a_n) to determine whether

$$\liminf_{n\to\infty} \frac{1}{a_n}\left(\min_{|x|=n} V(x) - \frac{1}{2}\ln n\right)$$

is 0 or $-\infty$, $\mathbf{P}^*$-a.s.

[N.B.: Since the preparation of the first draft of these notes, Questions 5.27
and 5.28 have been completely solved under Assumption (H) by Hu [131] in terms
of an integral test for $\min_{|x|=n} V(x)$; in particular, $\lim\inf_{n\to\infty} \frac{1}{\ln\ln n}(\min_{|x|=n} V(x) -$
$\frac{1}{2}\ln n) = -1$, $\mathbf{P}^*$-a.s.]

---

[10]Recall that $a_+ := \max\{a, 0\}$.

5.6 Convergence of the Additive Martingale

Assume $\psi(0) > 0$ and $\psi(1) = 0 = \psi'(1)$. The Biggins martingale convergence theorem (Theorem 3.2 in Sect. 3.2) tells us that under $\mathbf{P}^*(\cdot) := \mathbf{P}(\cdot \mid \text{non-extinction})$, we have $W_n \to 0$ almost surely. It is natural (see Biggins and Kyprianou [58]) to ask at which rate W_n goes to 0. This is usually called the Seneta–Heyde norming problem, referring to the Seneta–Heyde theorem for Galton–Watson processes [129, 218].

Our answer is as follows.

Theorem 5.29 *Under Assumption* (H), *we have, under* $\mathbf{P}^*$,

$$\lim_{n\to\infty} n^{1/2} W_n = \left(\frac{2}{\pi\sigma^2}\right)^{1/2} D_\infty, \quad \text{in probability,}$$

where $D_\infty > 0$, $\mathbf{P}^*$-*a.s., is the random variable in Theorem 5.2, and*

$$\sigma^2 := \mathbf{E}\left[\sum_{|x|=1} V(x)^2 e^{-V(x)}\right] \in (0, \infty).$$

One may wonder whether it is possible to strengthen convergence in probability in Theorem 5.29 into almost sure convergence. The answer is no.

Proposition 5.30 *Under Assumption* (H), *we have*

$$\limsup_{n\to\infty} n^{1/2} W_n = \infty, \quad \mathbf{P}^*\text{-a.s.}$$

Proof By definition, $W_n \geq \exp[-\inf_{|x|=n} V(x)]$. So the proposition follows immediately from (5.57) in Theorem 5.23 (Sect. 5.5). $\qquad\square$

Proposition 5.30 leads naturally to the following question.

Question 5.31 What is the rate at which the upper limits of $n^{1/2} W_n$ go to infinity $\mathbf{P}^*$-almost surely?

Questions 5.27 and 5.31 are clearly related via $W_n \geq \exp[-\inf_{|x|=n} V(x)]$. It is, however, not clear whether answering one of the questions will necessarily lead to answering the other.

Conjecture 5.32 Under Assumption (H), we have

$$\liminf_{n\to\infty} n^{1/2} W_n = \left(\frac{2}{\pi\sigma^2}\right)^{1/2} D_\infty, \quad \mathbf{P}^*\text{-a.s.,}$$

where $\sigma^2 := \mathbf{E}[\sum_{|x|=1} V(x)^2 e^{-V(x)}]$.

[N.B.: Again, since the preparation of the first draft of these notes, things have progressed: Question 5.31 and Conjecture 5.32 have been settled by Hu [131]. More precisely, Question 5.31 is completely solved in terms of an integral test, whereas the answer to Conjecture 5.32 is in the affirmative under an integrability condition which is slightly stronger than Assumption (H).]

5.7 The Genealogy of the Leftmost Position

We now look at the sample path of the branch in the branching random walk leading to the minimal position[11] at time n. Intuitively, it should behave like a Brownian motion on $[0, n]$, starting at 0 and ending around $\frac{3}{2} \ln n$, and staying above the line $i \mapsto \frac{\frac{3}{2} \ln n}{n} i$ for $0 \le i \le n$. If we normalise this sample path with the same scaling as Brownian motion, then we would expect it to behave asymptotically like a normalised Brownian excursion. This is rigorously proved by Chen [87].

More precisely, let $|m^{(n)}| = n$ be such that $V(m^{(n)}) = \min_{|x|=n} V(x)$, and for $0 \le i \le n$, let $m_i^{(n)}$ be the ancestor of $m^{(n)}$ in the i-th generation. Let $\sigma^2 := \mathbf{E}(\sum_{|x|=1} V(x)^2 e^{-V(x)})$ as before.

Recall that a normalised Brownian excursion can be formally defined as a standard Brownian bridge conditioned to be non-negative; rigorously, if $(B(t), t \ge 0)$ is a standard Brownian motion, writing $\mathfrak{g} := \sup\{t \le 1 : B(t) = 0\}$ and $\mathfrak{d} := \inf\{t \ge 1 : B(t) = 0\}$, then $(\frac{|B(\mathfrak{g}+(\mathfrak{d}-\mathfrak{g})t)|}{(\mathfrak{d}-\mathfrak{g})^{1/2}}, t \in [0, 1))$ is a normalised Brownian excursion.

Theorem 5.33 (Chen [87]) *Under Assumption (H),*

$$\left(\frac{V(m_{\lfloor nt \rfloor}^{(n)})}{(\sigma^2 n)^{1/2}}, t \in [0, 1] \right)$$

converges weakly to the normalised Brownian excursion, in $D([0, 1], \mathbb{R})$, the space of all càdlàg functions on $[0, 1]$ endowed with the Skorokhod topology.

For any vertex x with $|x| \ge 1$, let us write

$$\overline{V}(x) := \max_{1 \le i \le |x|} V(x_i), \tag{5.59}$$

---

[11]If there are several minima, one can, as before, choose any one at random according to the uniform distribution.

which stands for the maximum value of the branching random walk along the path connecting the root and x. How small can $\overline{V}(x)$ be when $|x| \to \infty$? If we take x to be a vertex on which the branching random walk reaches the minimum value at generation n, then we have seen in the previous paragraph that $\overline{V}(x)$ is of order of magnitude $n^{1/2}$. Can we do better?

The answer is yes. Recall that $\psi(t) := \ln \mathbf{E}(\sum_{|x|=1} e^{-tV(x)})$, $t \in \mathbb{R}$.

Theorem 5.34 (Fang and Zeitouni [107], Faraud et al. [111]) *Under Assumption* (H),

$$\lim_{n \to \infty} \frac{1}{n^{1/3}} \min_{|x|=n} \overline{V}(x) = \left(\frac{3\pi^2 \sigma^2}{2}\right)^{1/3}, \quad \mathbf{P}^*\text{-a.s.},$$

where $\mathbf{P}^*(\cdot) := \mathbf{P}(\cdot \mid \text{non-extinction})$ *as before.*

Theorem 5.34, which will be useful in Sect. 7.3, can be proved by means of a second-moment argument using the spinal decomposition theorem, using Mogulskii [197]'s small deviation estimates for sums of independent random variables. The theorem was originally proved in [107, 111] under stronger moment assumptions. For a proof under Assumption (H), see Mallein [188], who actually does not need the finiteness assumption of $\mathbf{E}[\widetilde{X} \ln_+ \widetilde{X}]$ in (5.3).

5.8 Proof of the Peeling Lemma

This section is devoted to the proof of the peeling lemma (Theorem 5.14 in Sect. 5.3). Fix $L \geq 0$ and $\varepsilon > 0$. Let $0 \leq K \leq z \leq \frac{3}{2} \ln n - L$.

Let $c_{75} = c_{75}(\varepsilon) > 0$ be the constant in Lemma A.6 of Appendix A.2. Recall the definition from (5.17): $\mathscr{L}_n^{z,L,K} = \{|x| = n : V(x_i) \geq a_i^{(n)}, 0 \leq i \leq n, V(x) \leq \frac{3}{2} \ln n - z + C\}$, where $a_i^{(n)}$ is defined in (5.16). Let $\beta_i^{(n)}$ be as in (5.19). By the many-to-one formula (Theorem 1.1 in Sect. 1.3),

$$\mathbf{Q}\left(w_n \in \mathscr{L}_n^{z,L,K}; \, \exists j \leq n-1, \, V(w_j) \leq a_{j+1}^{(n)} + 2\beta_{j+1}^{(n)} - c_{75}\right)$$

$$= \mathbf{P}\left(S_i \geq a_i^{(n)}, \, 0 \leq i \leq n; \, S_n \leq \frac{3}{2}\ln n - z + C; \right.$$

$$\left. \exists j \leq n-1, \, S_j \leq a_{j+1}^{(n)} + 2\beta_{j+1}^{(n)} - c_{75}\right),$$

which is bounded by $(L+1)^2 \varepsilon \frac{1+z-K}{n^{3/2}}$, for all sufficiently large n (Lemma A.6 of Appendix A.2). Since $(L+1)^2 \varepsilon$ can be as small as possible (L being fixed), it remains to show the existence of ρ such that

$$\mathbf{Q}\left(w_n \in \mathscr{L}_n^{z,L,K}; \, V(w_j) > a_{j+1}^{(n)} + 2\beta_{j+1}^{(n)} - c_{75}, \, 0 \leq j \leq n-1;\right.$$

$$\exists k \le n, \quad \sum_{y \in \mathrm{brot}(w_k)} [1 + (V(y) - a_k^{(n)})_+] e^{-V(y) + a_k^{(n)}} > \rho \, e^{-\beta_k^{(n)}} \Big)$$

$$\le \varepsilon \, \frac{1 + z - K}{n^{3/2}}.$$

On the event $\{V(w_{k-1}) > a_k^{(n)} + 2\beta_k^{(n)} - c_{75}\}$, we have

$$e^{-\beta_k^{(n)}} \ge e^{-c_{75}/2} \, e^{-[V(w_{k-1}) - a_k^{(n)}]/2}, \qquad (5.60)$$

and also

$$V(w_{k-1}) \ge a_k^{(n)}, \qquad (5.61)$$

if moreover $w_n \in \mathscr{L}_n^{z,L,K}$ (which guarantees $V(w_{k-1}) \ge a_{k-1}^{(n)}$, and $a_{k-1}^{(n)}$ differs from $a_{k-1}^{(n)}$ only if $k = \frac{n}{2}$, in which case $2\beta_k^{(n)} - c_{75} = 2(\frac{n}{2})^{1/7} - c_{75} \ge 0$ for $n \ge n_0$). On the other hand, by the elementary inequality $1 + (r + s)_+ \le (1 + r_+)(1 + s_+)$ for all $r, s \in \mathbb{R}$, we have

$$\sum_{y \in \mathrm{brot}(w_k)} [1 + (V(y) - a_k^{(n)})_+] e^{-V(y) + a_k^{(n)}}$$

$$\le [1 + (V(w_{k-1}) - a_k^{(n)})_+] e^{-V(w_{k-1}) + a_k^{(n)}} \Lambda(w_k),$$

where, for all $x \in \mathbb{T}$ with $|x| \ge 1$,

$$\Lambda(x) := \sum_{y \in \mathrm{brot}(x)} [1 + (\Delta V(y))_+] e^{-\Delta V(y)},$$

and $\Delta V(y) := V(y) - V(\overleftarrow{y})$ (recalling that $\overleftarrow{y}$ is the parent of y). Therefore, on the event $\{\sum_{y \in \mathrm{brot}(w_k)} [1 + (V(y) - a_k^{(n)})_+] e^{-V(y) + a_k^{(n)}} > \rho \, e^{-\beta_k^{(n)}}\} \cap \{V(w_k) > a_{k+1}^{(n)} + 2\beta_{k+1}^{(n)} - c_{75}\}$, we have

$$\rho \, e^{-c_{75}/2} e^{-[V(w_{k-1}) - a_k^{(n)}]/2} \le \rho \, e^{-\beta_k^{(n)}} \quad \text{(by (5.60))}$$

$$< [1 + (V(w_{k-1}) - a_k^{(n)})_+] e^{-[V(w_{k-1}) - a_k^{(n)}]} \Lambda(w_k),$$

which implies that (writing $c_{43} := \rho \, e^{-c_{75}/2}$)

$$\Lambda(w_k) \ge c_{43} \, \frac{e^{[V(w_{k-1}) - a_k^{(n)}]/2}}{1 + (V(w_{k-1}) - a_k^{(n)})_+} \ge c_{44} c_{43} \, e^{[V(w_{k-1}) - a_k^{(n)}]/3},$$

where $c_{44} := \inf_{x \geq 0} \frac{e^{x/6}}{1+x} > 0$. The inequality can be rewritten as $V(w_{k-1}) \leq a_k^{(n)} + 3\ln\frac{\Lambda(w_k)}{\rho_1}$, where $\rho_1 := c_{44}c_{43} = c_{44}\,e^{-c_{75}/2}\rho$. Since $V(w_{k-1}) \geq a_k^{(n)}$ (see (5.61)), we also have $\Lambda(w_k) \geq \rho_1$.

The proof of the lemma is reduced to showing the following: there exists $\rho > 0$ such that with $\rho_1 := c_{44}\,e^{-c_{75}/2}\rho$,

$$\sum_{k=1}^{n} \mathbf{Q}\Big(w_n \in \mathscr{L}_n^{z,L,K};\ A_{k,n}\Big) \leq \varepsilon\,\frac{1+z-K}{n^{3/2}}, \tag{5.62}$$

where

$$A_{k,n} := \Big\{V(w_{k-1}) \leq a_k^{(n)} + 3\ln\frac{\Lambda(w_k)}{\rho_1},\ \Lambda(w_k) \geq \rho_1\Big\}.$$

We decompose $\sum_{k=1}^{n}$ into the sum of $\sum_{k=1}^{\frac{3}{4}n}$ and $\sum_{k=\frac{3}{4}n+1}^{n}$, and prove that both are bounded by $\varepsilon\,\frac{1+z-K}{n^{3/2}}$ (so at the end, we should replace ε by $\frac{\varepsilon}{2}$).

First Situation: $1 \leq k \leq \frac{3}{4}n$. We have

$$\mathbf{Q}(w_n \in \mathscr{L}_n^{z,L,K};\ A_{k,n}) = \mathbf{E}_{\mathbf{Q}}\Big[\psi_{k,n}(V(w_k))\,\mathbf{1}_{\{A_{k,n};\ V(w_i)\geq a_i^{(n)},\ 0\leq i\leq k\}}\Big],$$

where

$$\psi_{k,n}(r) := \mathbf{Q}_r\Big(V(w_j) \geq a_{k+j}^{(n)},\ 0\leq j\leq n-k;\ V(w_{n-k}) \leq \frac{3}{2}\ln n - z + C\Big)$$
$$= \mathbf{P}_r\Big(S_j \geq a_{k+j}^{(n)},\ 0\leq j\leq n-k;\ S_{n-k} \leq \frac{3}{2}\ln n - z + C\Big),$$

where the last identity follows from the many-to-one formula (Theorem 1.1 in Sect. 1.3).

Since $k \leq \frac{3}{4}n$, $\psi_{k,n}(r) \leq c_{45}\frac{1+(r-a_k^{(n)})_+}{n^{3/2}}$ with $c_{45} = c_{45}(L,C)$ (by applying Lemma A.1 of Appendix A.2 if $\frac{n}{2} < k \leq \frac{3}{4}n$, and Lemma A.4 if $k \leq \frac{n}{2}$; noting that the probability expression in Lemma A.4 is non-decreasing in λ, so the lemma applies even if $k \leq \frac{n}{2}$ is close to $\frac{n}{2}$). Hence

$$\sum_{k=1}^{\frac{3}{4}n} \mathbf{Q}(w_n \in \mathscr{L}_n^{z,L,K};\ A_{k,n})$$

$$\leq \frac{c_{45}}{n^{3/2}} \sum_{k=1}^{\frac{3}{4}n} \mathbf{E}_{\mathbf{Q}}\Big[[1+(V(w_k)-a_k^{(n)})_+]\,\mathbf{1}_{\{A_{k,n};\ V(w_i)\geq a_i^{(n)},\ 0\leq i\leq k\}}\Big].$$

Note that $1 + (V(w_k) - a_k^{(n)})_+ \leq [1 + (V(w_{k-1}) - a_k^{(n)})_+] + (\Delta V(w_k))_+$, where $\Delta V(y) := V(y) - V(\overset{\leftarrow}{y})$ as before. The proof of $\sum_{k=1}^{\frac{3}{4}n} \mathbf{Q}(A_{k,n}) \leq \varepsilon \frac{1+z-K}{n^{3/2}}$ will be complete once we establish the following estimates: there exists $\rho > 0$ such that with $\rho_1 := c_{44} e^{-c_{75}/2} \rho$,

$$\sum_{k=1}^{\frac{3}{4}n} [I_{(5.63)}(k) + I\!I_{(5.63)}(k)] \leq \varepsilon (1 + z - K), \tag{5.63}$$

where

$$I_{(5.63)}(k) := \mathbf{E}_\mathbf{Q}\left[[1 + (V(w_{k-1}) - a_k^{(n)})_+] \mathbf{1}_{\{A_{k,n};\, V(w_i) \geq a_i^{(n)},\, 0 \leq i \leq k\}}\right],$$

$$I\!I_{(5.63)}(k) := \mathbf{E}_\mathbf{Q}\left[(\Delta V(w_k))_+ \mathbf{1}_{\{A_{k,n};\, V(w_i) \geq a_i^{(n)},\, 0 \leq i \leq k\}}\right].$$

Recall that $A_{k,n} := \{V(w_{k-1}) \leq a_k^{(n)} + 3 \ln \frac{\Lambda(w_k)}{\rho_1},\ \Lambda(w_k) \geq \rho_1\}$. By the spinal decomposition theorem, $(V(w_i),\ 0 \leq i < k)$ is independent of $(\Lambda(w_k),\ \Delta V(w_k))$.

We first study $I_{(5.63)}(k)$. On $A_{k,n}$, we have $V(w_{k-1}) - a_k^{(n)} \leq 3 \ln \frac{\Lambda(w_k)}{\rho_1}$. We condition on $\Lambda(w_k)$, and observe that uniformly in $r \geq \rho_1$,

$$\sum_{k=1}^{\frac{n}{2}} \mathbf{Q}\left(V(w_{k-1}) \leq a_k^{(n)} + 3 \ln \frac{r}{\rho_1};\ V(w_i) \geq a_i^{(n)},\ 0 \leq i \leq k\right)$$

$$= \sum_{k=1}^{\frac{n}{2}} \mathbf{Q}\left(V(w_{k-1}) \leq -z + K + 3 \ln \frac{r}{\rho_1};\ V(w_i) \geq -z + K,\ 0 \leq i \leq k\right)$$

$$\leq \sum_{k=1}^{\infty} \mathbf{P}_{z-K}\left(S_{k-1} \leq 3 \ln \frac{r}{\rho_1};\ S_i \geq 0,\ 0 \leq i \leq k-1\right),$$

which is bounded by $c_{46} (1 + z - K)(1 + \ln \frac{r}{\rho_1})$ according to Lemma A.5 of Appendix A.2. On the other hand, for $r \geq \rho_1$ and $\frac{n}{2} < k \leq \frac{3}{4}n$,

$$\mathbf{Q}\left(V(w_{k-1}) \leq a_k^{(n)} + 3 \ln \frac{r}{\rho_1};\ V(w_i) \geq a_i^{(n)},\ 0 \leq i \leq k\right)$$

$$= \mathbf{P}_{z-K}\left(S_{k-1} \leq \frac{3}{2} \ln n - K - L + 3 \ln \frac{r}{\rho_1};\ S_i \geq 0,\ 0 \leq i \leq \frac{n}{2},\right.$$

$$\left. S_j \geq \frac{3}{2} \ln n - K - L,\ \frac{n}{2} < j \leq k\right),$$

which, by the Markov property at time $\frac{n}{2}$, is

$$\mathbf{E}_{z-K}\left[\mathbf{1}_{\{S_i \geq 0,\, 0 \leq i \leq \frac{n}{2}\}} f_{(5.64)}\left(S_{\frac{n}{2}}\right)\right],$$

where, for $u \geq 0$,

$$f_{(5.64)}(u) := \mathbf{P}_u\Big(S_{k-1-\frac{n}{2}} \leq \frac{3}{2}\ln n - K - L + 3\ln\frac{r}{\rho_1};$$

$$S_\ell \geq \frac{3}{2}\ln n - K - L,\ 0 < \ell \leq k - \frac{n}{2}\Big). \qquad (5.64)$$

By Lemma A.5 of Appendix A.2, for $u \geq \frac{3}{2}\ln n - K - L$ and $r \geq \rho_1$,

$$\sum_{k=\frac{n}{2}+1}^{\frac{3}{4}n} f_{(5.64)}(u) \leq c_{47}\,(1 + 3\ln\frac{r}{\rho_1})[1 + (u - \frac{3}{2}\ln n + K + L)].$$

Consequently, for $z \leq \frac{3}{2}\ln n - L$,

$$\sum_{k=\frac{n}{2}+1}^{\frac{3}{4}n} \mathbf{Q}\Big(V(w_{k-1}) \leq a_k^{(n)} + 3\ln\frac{r}{\rho_1};\ V(w_i) \geq a_i^{(n)},\ 0 \leq i \leq k\Big)$$

$$\leq c_{47}\,(1 + 3\ln\frac{r}{\rho_1})\,\mathbf{E}_{z-K}\left[\mathbf{1}_{\{S_i \geq 0,\, 0 \leq i \leq \frac{n}{2}\}}\,[1 + (S_{\frac{n}{2}} - \frac{3}{2}\ln n + K + L)_+]\right]$$

$$\leq c_{47}\,(1 + 3\ln\frac{r}{\rho_1})\,\mathbf{E}\left[\mathbf{1}_{\{S_i \geq -z+K,\, 0 \leq i \leq \frac{n}{2}\}}\,[1 + (S_{\frac{n}{2}})_+]\right],$$

which is bounded by $c_{48}\,(1 + 3\ln\frac{r}{\rho_1})(1 + z - K)$ according to Lemma A.3 of Appendix A.2.

Summarizing, we have

$$\sum_{k=1}^{\frac{3}{4}n} I_{(5.63)}(k) \leq c_{49}\,(1 + z - K)\,\mathbf{E_Q}\left[(1 + \ln\frac{\Lambda(w_k)}{\rho_1})^2\,\mathbf{1}_{\{\Lambda(w_k) \geq \rho_1\}}\right]$$

$$\leq c_{49}\,(1 + z - K)\,\mathbf{E_Q}\left[(1 + \ln_+ \Lambda(w_k))^2\,\mathbf{1}_{\{\Lambda(w_k) \geq \rho_1\}}\right],$$

if $\rho_1 \geq 1$. Since $\Lambda(w_k)$ is distributed as $\Lambda(w_1)$ (under $\mathbf{Q}$), which is bounded by $X + \widetilde{X}$ with the notation of (5.4) at the beginning of the chapter, we have

$$\sum_{k=1}^{\frac{3}{4}n} I_{(5.63)}(k) \leq c_{49}\,(1 + z - K)\,\mathbf{E_Q}\left[(1 + \ln_+(X + \widetilde{X}))^2\,\mathbf{1}_{\{X + \widetilde{X} \geq \rho_1\}}\right]$$

$$= c_{49} \left(1 + z - K\right) \mathbf{E}\left[X(1 + \ln_+(X + \widetilde{X}))^2 \, \mathbf{1}_{\{X + \widetilde{X} \geq \rho_1\}}\right]$$

$$\leq \varepsilon \left(1 + z - K\right), \tag{5.65}$$

if we choose $\rho_1 := c_{44}\, e^{-c_{75}/2}\, \rho$ sufficiently large (because $\mathbf{E}[X(1 + \ln_+(X + \widetilde{X}))^2] < \infty$; see (5.3) and (5.5)), i.e., if we choose ρ sufficiently large.

We now turn to $I\!I_{(5.63)}(k)$, and use the same argument, except that we condition on the pair $(\Delta V(w_k),\, \Lambda(w_k))$, to see that

$$\sum_{\ell=1}^{\frac{3}{4}n} I\!I_{(5.63)}(k) \leq c_{50} \left(1 + z - K\right) \mathbf{E}_{\mathbf{Q}}\left[(\Delta(w_k))_+ (1 + \ln_+ \Lambda(w_k)) \, \mathbf{1}_{\{\Lambda(w_k) \geq \rho_1\}}\right].$$

The random vector $(\Delta(w_k),\, \Lambda(w_k))$ under $\mathbf{Q}$ is distributed as $(\Delta(w_1),\, \Lambda(w_1))$. We have noted that $\Lambda(w_1) \leq X + \widetilde{X}$; on the other hand, $\Delta(w_1) = V(w_1)$. Since $\mathbf{Q}(w_1 = x \mid \mathscr{F}_1) = \frac{e^{-V(x)}}{X}$ for any $x \in \mathbb{T}$ with $|x| = 1$, where $\mathscr{F}_1$ is the σ-field generated by the first generation of the branching random walk, we have $\mathbf{E}_{\mathbf{Q}}(V(w_1)_+ \mid \mathscr{F}_1) = \frac{1}{X} \sum_{|x|=1} V(x)_+ e^{-V(x)} = \frac{\widetilde{X}}{X}$; hence

$$\mathbf{E}_{\mathbf{Q}}\left[(\Delta(w_k))_+ (1 + \ln_+ \Lambda(w_k)) \, \mathbf{1}_{\{\Lambda(w_k) \geq \rho_1\}}\right]$$

$$\leq \mathbf{E}_{\mathbf{Q}}\left[V(w_1)_+ (1 + \ln_+(X + \widetilde{X})) \, \mathbf{1}_{\{X + \widetilde{X} \geq \rho_1\}}\right]$$

$$= \mathbf{E}_{\mathbf{Q}}\left[\frac{\widetilde{X}}{X} (1 + \ln_+(X + \widetilde{X})) \, \mathbf{1}_{\{X + \widetilde{X} \geq \rho_1\}}\right]$$

$$= \mathbf{E}\left[\widetilde{X} (1 + \ln_+(X + \widetilde{X})) \, \mathbf{1}_{\{X + \widetilde{X} \geq \rho_1\}}\right],$$

which is smaller than $\frac{\varepsilon}{c_{50}}$ if $\rho_1 := c_{44}\, e^{-c_{75}/2}\, \rho$ is sufficiently large, since $\mathbf{E}[\widetilde{X}(1 + \ln_+(X + \widetilde{X}))] < \infty$ (again by (5.3) and (5.5)). As a consequence, if the constant ρ is chosen sufficiently large, then

$$\sum_{k=1}^{\frac{3}{4}n} I\!I_{(5.63)}(k) \leq \varepsilon \left(1 + z - K\right),$$

which, combined with (5.65), yields $\sum_{k=1}^{\frac{3}{4}n} \mathbf{Q}(w_n \in \mathscr{L}_n^{z,L,K};\, A_{k,n}) \leq \varepsilon\, \frac{1+z-K}{n^{3/2}}$.

Second (and Last) Situation: $\frac{3}{4}n < k \leq n$. For notational simplification, we write

$$V_j := V(w_j),$$

$$U_j := V(w_n) - V(w_{n-j}) = V_n - V_{n-j},$$

$$\Lambda_j := \Lambda(w_j), \quad 0 \leq j \leq n.$$

By the spinal decomposition theorem, $(U_j, \ 1 \le j \le n - k)$ and Λ_k are independent (under $\mathbf{Q}$), and so are $(U_j, \ 1 \le j \le n - k; \ U_{n-k+1}; \ \Lambda_k)$ and $(V_i, \ 0 \le i \le k - 1)$.

We use some elementary argument. Let $\frac{3}{4}n < k \le n$. By definition, $a_k^{(n)} = a_n^{(n)} = \frac{3}{2} \ln n - z - L$, so on the event $\{w_n \in \mathscr{L}_n^{z,L,K}\}$, we have $V_{k-1} - a_k^{(n)} = (V_n - U_{n-k+1}) - a_k^{(n)} \ge (a_n^{(n)} - U_{n-k+1}) - a_k^{(n)} = -U_{n-k+1}$, which yields $\Lambda_k \ge \rho_1 \, e^{-U_{n-k+1}/3}$. Also, on the event $\{w_n \in \mathscr{L}_n^{z,L,K}\}$, for $0 \le j < \frac{n}{2}$ (so that $a_{n-j}^{(n)} = \frac{3}{2} \ln n - z - L$), $U_j = V_n - V_{n-j} \le (\frac{3}{2} \ln n - z + C) - a_{n-j}^{(n)} = L + C$. Moreover, since $V(w_n) = V_{k-1} + U_{n-k+1}$, the condition $\frac{3}{2} \ln n - z - L \le V(w_n) \le \frac{3}{2} \ln n - z + C$ on the event $\{w_n \in \mathscr{L}_n^{z,L,K}\}$ reads $\frac{3}{2} \ln n - z - L - U_{n-k+1} \le V_{k-1} \le \frac{3}{2} \ln n - z + C - U_{n-k+1}$. As such, we have

$$\left(w_n \in \mathscr{L}_n^{z,L,K}; \ A_{k,n} \right) \subset A_{k,n}^{(1)} \cap A_{k,n}^{(2)},$$

where

$$A_{k,n}^{(1)} := \{ V_i \ge a_i^{(n)}, \ 0 \le i \le k - 1;$$
$$\frac{3}{2} \ln n - z - L - U_{n-k+1} \le V_{k-1} \le \frac{3}{2} \ln n - z + C - U_{n-k+1} \},$$

$$A_{k,n}^{(2)} := \{ \Lambda_k \ge \rho_1 \, (e^{-U_{n-k+1}/3} \vee 1); \ U_j \le L + C, \ 0 \le j < \frac{n}{2} \}.$$

[We have already noted that the inequality $\Lambda_k \ge \rho_1$ holds trivially on $A_{k,n}$.] Since $\frac{3}{4}n < k \le n$, we have $\frac{n}{2} \ge n - k + 1$, so $A_{k,n}^{(2)} \subset A_{k,n}^{(3)}$, where

$$A_{k,n}^{(3)} := \{ \Lambda_k \ge \rho_1 \, (e^{-U_{n-k+1}/3} \vee 1); \ U_j \le L + C, \ 0 \le j < n - k + 1 \}.$$

Let $\mathscr{G}_{k,n} := \sigma(U_j, \ 1 \le j \le n - k; \ U_{n-k+1}; \ \Lambda_k)$. We have already observed that $(V_i, \ 0 \le i \le k - 1)$ is independent of $\mathscr{G}_{k,n}$. Since $A_{k,n}^{(3)} \in \mathscr{G}_{k,n}$, this yields

$$\mathbf{Q}(w_n \in \mathscr{L}_n^{z,L,K}; \ A_{k,n} \,|\, \mathscr{G}_{k,n}) \le \mathbf{1}_{A_{k,n}^{(3)}} \, \mathbf{Q}(A_{k,n}^{(1)} \,|\, \mathscr{G}_{k,n}) = \mathbf{1}_{A_{k,n}^{(3)}} \, f_{k,n}(U_{n-k+1}),$$

where, for $r \le L + C$,

$$f_{k,n}(r) := \mathbf{Q}\Big(V_i \ge a_i^{(n)}, \ 0 \le i \le k - 1;$$
$$\frac{3}{2} \ln n - z - L - r \le V_{k-1} \le \frac{3}{2} \ln n - z + C - r \Big).$$

By the spinal decomposition theorem, $(V_i, \ 0 \le i \le k - 1)$ under $\mathbf{Q}$ is distributed as $(S_i, \ 0 \le i \le k-1)$ under $\mathbf{P}$, so we are entitled to apply Lemma A.4 of Appendix A.2

to see that for $r \leq L + C$,

$$f_{k,n}(r) \leq c_{51} \frac{(L + C + 1)(z - K + L + C + 1)(2L + 2C - r + 1)}{(k - 1)^{3/2}}$$

$$\leq c_{52} \frac{(z - K + 1)(L + C - r + 1)}{n^{3/2}},$$

with $c_{52} = c_{52}(L, C)$. Consequently,

$$\mathbf{Q}(w_n \in \mathscr{L}_n^{z,L,K}; A_{k,n}) \leq c_{52} \frac{z - K + 1}{n^{3/2}} \mathbf{E_Q}\left[(L + C - U_{n-k+1} + 1) \mathbf{1}_{A_{k,n}^{(3)}}\right].$$

On the event $A_{k,n}^{(3)}$, we have $\Lambda_k \geq \rho_1 \, e^{-U_{n-k+1}/3} = \rho_1 \, e^{-(U_{n-k}+\Delta_k)/3}$, where $\Delta_k = \Delta(w_k) := V(w_k) - V(w_{k-1})$; so $-U_{n-k} \leq \Delta_k + 3 \ln \frac{\Lambda_k}{\rho_1} \leq (\Delta_k)_+ + 3 \ln \frac{\Lambda_k}{\rho_1}$, whereas $-U_{n-k+1} \leq 3 \ln \frac{\Lambda_k}{\rho_1}$. Therefore,

$$\mathbf{E_Q}\left[(L + C - U_{n-k+1} + 1) \mathbf{1}_{A_{k,n}^{(3)}}\right] \leq \mathbf{E_Q}\left[(L + C + 3 \ln \frac{\Lambda_k}{\rho_1} + 1) \mathbf{1}_{\{\Lambda_k \geq \rho_1\}}\right.$$

$$\left. \times \mathbf{1}_{\{-U_{n-k} \leq (\Delta_k)_+ + 3 \ln \frac{\Lambda_k}{\rho_1}\}} \mathbf{1}_{\{U_j \leq L+C, 0 \leq j \leq n-k\}}\right].$$

On the right-hand side, $(U_j, \; 0 \leq j \leq n-k)$ is independent of (Λ_k, Δ_k) by the spinal decomposition. By Lemma A.5 of Appendix A.2 (applied to $(-S_i)$, which is also a centred random walk),

$$\sum_{k=\frac{3}{4}n+1}^{n} \mathbf{Q}\left(-U_{n-k} \leq x, \; -U_j \geq -(L + C), \; 0 \leq j \leq n - k\right) \leq c_{53} (x + 1),$$

for some constant $c_{53} = c_{53}(L, C) > 0$, all $x \geq 0$ and $n \geq 1$. Consequently,

$$\sum_{k=\frac{3}{4}n+1}^{n} \mathbf{E_Q}\left[(L + C - U_{n-k+1} + 1) \mathbf{1}_{A_{k,n}^{(3)}}\right]$$

$$\leq c_{53} \, \mathbf{E_Q}\left[(L + C + 3 \ln \frac{\Lambda_k}{\rho_1} + 1) \, ((\Delta_k)_+ + 3 \ln \frac{\Lambda_k}{\rho_1} + 1) \mathbf{1}_{\{\Lambda_k \geq \rho_1\}}\right]$$

$$\leq c_{54} \, \mathbf{E_Q}\left[(1 + \ln_+ \Lambda_k) \, ((\Delta_k)_+ + \ln_+ \Lambda_k + 1) \mathbf{1}_{\{\Lambda_k \geq \rho_1\}}\right],$$

for some $c_{54} = c_{54}(L, C)$ if $\rho_1 \geq 1$. We have already seen that it is possible to choose ρ_1 sufficiently large such that the term $\mathbf{E_Q}[\cdots]$, which does not depend on k, is bounded by $\frac{\varepsilon}{c_{54}c_{52}}$. As a consequence,

$$\sum_{k=\frac{3}{4}n+1}^{n} \mathbf{Q}(w_n \in \mathscr{L}_n^{z,L,K}; A_{k,n}) \le \varepsilon \frac{z-K+1}{n^{3/2}},$$

as desired. This completes the proof of the peeling lemma.   □

5.9 Notes

The existence of $t^* > 0$ satisfying (5.1) in Sect. 5.1 is a basic assumption in obtaining the universality results presented in this chapter. In the literature, discussions on (5.1) are spread in a few places; see Comets [89] (quoted in Mörters and Ortgiese [198]), the appendices in Jaffuel [146] (in the ArXiv version, but not in the published version) and Bérard and Gouéré [41]. When (5.1) has no solution, the asymptotic behaviour of the branching random walk depends strongly on the distribution of the governing point process Ξ; for study of the leftmost position, see Bramson [68], Dekking and Host [90], Amini et al. [23].

Theorem 5.2(i) in Sect. 5.2 is due to Biggins and Kyprianou [57]. Theorem 5.2(ii) and Lemma 5.5, first proved by Biggins and Kyprianou [57] under slightly stronger assumptions, are borrowed from Aïdékon [8]. The idea used in the proof of Lemma 5.5 goes back to Lyons [173] in his elegant proof (Sect. 4.8) of the Biggins martingale convergence theorem.

Theorem 5.12 in Sect. 5.3 is proved independently by Hu and Shi [137] and by Addario-Berry and Reed [3], both under stronger conditions, while a shorter proof is presented in [14]. Addario-Berry and Reed [3] also prove an analogous result for the mean of the minimal position, and obtain, moreover, an exponential tail estimate for the difference between the minimal position and its mean. Precise moderate deviation probabilities are obtained by Hu [132]. Previous important work includes the slow $\ln n$ rate appearing in McDiarmid [179] who, under additional assumptions, obtains the upper bound in Theorem 5.12 without getting the optimal value $3/2$, and tightness in Bramson and Zeitouni [70]; also, convergence in a special case (including the example of Gaussian displacements), centered at the mean, is proved by Bachmann [33].

The proof of Theorem 5.15 in Sect. 5.4 is adapted from Aïdékon [8], with some simplification which I have learned from Chen [87]; in particular, by placing the absorbing barrier starting at level $-z + K$ instead of at the origin, we do not need to study the number of negative excursions. This simplification is also found in the recent work of Bramson et al. [71], where a more compact proof is presented, with a particularly interesting new ingredient: instead of counting the number of certain vertices at generation n as we do in (5.35), the authors count at generation $n-\ell$, with ℓ independent of n but going to infinity after the passage $n \to \infty$. This allows for a nice-looking law of large numbers instead of the fraction in (5.35). Also, discussions are provided on the lattice case.

After the publication of [8], a new version has been added by the author on ArXiv with some simple modifications so that Theorem 5.15 remains valid even when #$\varXi$ is possibly infinite.

Weak convergence for the minimum of general log-correlated Gaussian fields (in any dimension), centered at the mean, is established by Ding et al. [96].

The idea of the first step (Sect. 5.4.1) in the proof of Theorem 5.15, expressing that it suffices to study tail behaviour instead of weak convergence, is already used in the study of branching Brownian motion by Bramson [67, 69]. This idea can be exploited in other problems (see for example an important application in the recent work of Arguin et al. [27] in the proof of weak convergence of positions in a branching random walk viewed from the leftmost particle, and the lecture notes of Berestycki [43] for several other applications). When (5.1) fails (so that Theorem 5.15 does not apply), the limiting law of the leftmost position is studied by Barral et al. [34].

The analogue of Theorem 5.15 for branching Brownian motion was previously known to Lalley and Sellke [164] (and is recalled in (1.3) of Sect. 1.1), while the convergence in distribution of $M_n - (3\ln n)/2$, for branching Brownian motion, to a travelling-wave solution of the KPP equation, is proved in the celebrated work of Bramson [69] (a fact which is also recalled in Sect. 1.1).

The original proof of Theorem 5.23 (Sect. 5.5) in [137] requires some stronger integrability conditions. The lim inf part, (5.57), as well as Theorem 5.24, are from [15]. For the analogous results for branching Brownian motion, see Roberts [215].

Theorem 5.29 (Sect. 5.6) can be found in [15]; it improves a previous result of [137]. For the particular model of i.i.d. Gaussian random variables on the edges of rooted regular trees, this is also known to Webb [233]. Convergence of the additive martingale in the near-critical regime is studied by Alberts and Ortgiese [18] and Madaule [182].

The statement of the peeling lemma, slightly more general than in Aïdékon [8] (where only the case $K = 0$ is studied because the absorbing barrier is placed in a different place), is borrowed from Chen [87]. Our proof, presented in Sect. 5.8, follows closely the arguments in [8], except that no family of Palm measures is used.

There is a big number of recent results on the extremal process in the branching random walks: it converges weakly to a limiting process which is a decorated Poisson point processes. See Arguin et al. [27–30], Aïdékon et al. [16], Maillard [184] for branching Brownian motion; Madaule [181] for branching random walks. See also Gouéré [117] for an analysis based on [16, 27, 28, 30], and Subag and Zeitouni [225] for a general view of decorated Poisson point processes. Several predictions by Brunet and Derrida [75, 76] concerning the limiting decorated Poisson point process (notably about spacings and distribution at infinity) for the branching Brownian motion still need to be proved rigorously. For a "spatial" version of convergence of the extremal process, see Bovier and Hartung [64]. For the study of extremes in models with a time-inhomogeneous branching mechanism, see Fang and Zeitouni [108, 109], Maillard and Zeitouni [187], Mallein [189, 190], Bovier and Hartung [63, 65].

Chapter 6
Branching Random Walks with Selection

We have studied so far various asymptotic properties of the branching random walk by means of the spinal decomposition theorem. We are now facing at two very short chapters where the branching random walk intervenes in more complicated models; these topics are close to my current research work. No proof is given, though most of the ingredients needed in the proofs have already been seen by us in the previous chapters.

The present chapter is devoted to a few models of branching random walks in presence of certain selection criteria.

6.1 Branching Random Walks with Absorption

Branching processes were introduced by Galton and Watson in the study of survival probability for families in Great Britain. In the supercritical case of the Galton–Watson branching process, when the system survives, the number of individuals in the population grows exponentially fast, a phenomenon that is not quite realistic in biology. From this point of view, it sounds natural to impose a criterion of *selection*, according to which only some individuals in the population are allowed to survive, while others (as well as their descendants) are eliminated from the system.

In this section, we consider branching random walks in the presence of an absorbing barrier: any individual lying above the barrier gets erased (= absorbed).

Let $(V(x))$ denote a supercritical branching random walk, i.e., $\psi(0) > 0$, where $\psi(t) := \ln \mathbf{E}(\sum_{|x|=1} e^{-tV(x)})$, $t \in \mathbb{R}$. The governing point process is still denoted by Ξ. We think of $V(x)$ as representing the weakness of the individual x: whenever the value of $V(x)$ goes above a certain slope, the individual x is removed from the system. Throughout the section, we assume $\psi(1) = 0 = \psi'(1)$, i.e., $\mathbf{E}(\sum_{|x|=1} e^{-V(x)}) = 1$ and $\mathbf{E}(\sum_{|x|=1} V(x)e^{-V(x)}) = 0$.

© Springer International Publishing Switzerland 2015
Z. Shi, *Branching Random Walks*, Lecture Notes in Mathematics 2151,
DOI 10.1007/978-3-319-25372-5_6

Let $\varepsilon \in \mathbb{R}$ denote the slope of the absorbing barrier. So the individual x gets removed whenever $V(x) > \varepsilon|x|$ (or whenever one of its ancestors has been removed). Recall that an infinite ray (x_i) is a sequence of vertices such that $x_0 := \varnothing < x_1 < x_2 < \ldots$ with $|x_i| = i$, $i \geq 0$. Let $p_{\text{surv}}(\varepsilon)$ denote the survival probability, i.e., the probability that there exists an infinite ray (x_i) such that $V(x_i) \leq \varepsilon i$ for all $i \geq 0$.

Theorem 6.1 (Biggins et al. [59]) *Under Assumption* (H), $p_{\text{surv}}(\varepsilon) > 0$ *if* $\varepsilon > 0$, *and* $p_{\text{surv}}(\varepsilon) = 0$ *if* $\varepsilon \leq 0$.

Proof The case $\varepsilon > 0$ follows from Lemma 5.13 of Sect. 5.3, whereas the case $\varepsilon \leq 0$ from Remark 5.21 of Sect. 5.4.3.   □

Since Assumption (H) implies that $\frac{1}{n} \min_{|x|=n} V(x) \to 0$ almost surely on the set of non-extinction, the statement of Theorem 6.1 is of no surprise.

It is not hard to see $p_{\text{surv}}(\varepsilon) \to 0$ when ε tends to 0, so it looks natural to ask its rate of decay. This is a question raised by Pemantle [207].

Theorem 6.2 (Gantert et al. [115]) *Assume* $\psi(0) > 0$ *and* $\psi(1) = 0 = \psi'(1)$. *If* $\psi(1+\delta) < \infty$, $\psi(-\delta) < \infty$ *and* $\mathbf{E}[(\#\varXi)^{1+\delta}] < \infty$ *for some* $\delta > 0$, *then*

$$p_{\text{surv}}(\varepsilon) = \exp\left(-(1+o(1))\frac{\pi\sigma}{(2\varepsilon)^{1/2}}\right), \quad \varepsilon \downarrow 0, \tag{6.1}$$

where $\sigma^2 := \mathbf{E}(\sum_{|x|=1} V(x)^2 e^{-V(x)})$.

The proof of Theorem 6.2 relies on a second-moment argument by applying the spinal decomposition theorem; see [115]. The assumption of finiteness of $\psi(1+\delta)$, $\psi(-\delta)$ and $\mathbf{E}[(\#\varXi)^{1+\delta}]$ for some $\delta > 0$ is not necessary; in fact, the truncating procedure in the second-moment argument used in [115] is not optimal. It is, however, not clear whether Assumption (H) suffices for the validity of Theorem 6.2. The same remark applies to most of the theorems in this chapter.

Theorem 6.2 also plays a crucial role in the study of branching random walks with competition, briefly described in the next section.

We now turn to the case $\varepsilon = 0$. Theorem 6.1 tells us that in this case, there is extinction of the system; only finitely many individuals appear in the system. What can be said of the total number of individuals in the system? This question was originally raised by Aldous [19].

More precisely, let

$$\mathscr{Z} := \{x \in \mathbb{T}: \ V(y) \leq 0, \ \forall y \in [\![\varnothing, x]\!]\},$$

where, as before, $[\![\varnothing, x]\!]$ is the shortest path on the tree connecting x to the root $\varnothing$.

It is conjectured by Aldous [19] that under suitable integrability assumptions, one would have $\mathbf{E}(\#\mathscr{Z}) < \infty$ but $\mathbf{E}[(\#\mathscr{Z}) \ln_+^2(\#\mathscr{Z})] = \infty$.

The conjecture is proved by Addario-Berry and Broutin [1], while Aïdékon et al. in [17], improving a previous result of Aïdékon [7], give the precise tail probability of $\#\mathscr{L}$. Under the assumption $\psi(1) = 0$, we can define the associated one-dimensional random walk (S_n) as in (4.1) of Sect. 4.2; let R be the renewal function of (S_n), as in (A.2) of Appendix A.1. Recall that Ξ is the point process governing the law of the branching random walk, and that under $\mathbf{P}_u$ (for $u \in \mathbb{R}$), the branching random walk starts with a particle at position u.

Theorem 6.3 (Aïdékon et al. [17]) *Assume $\psi(0) > 1$ and $\psi(1) = 0 = \psi'(1)$. If for some $\delta > 0$, $\psi(1 + \delta) < \infty$, $\psi(-\delta) < \infty$ and $\mathbf{E}[(\#\Xi)^{2+\delta}] < \infty$, then there exists a constant $c \in (0, \infty)$ such that for any $u \geq 0$,*

$$\mathbf{P}_u(\#\mathscr{L} > n) \sim c\, \frac{R(u)}{n(\ln n)^2}, \quad n \to \infty.$$

Let us have another look at Theorem 6.1, which tells us that the slope $\varepsilon = 0$ is critical in some sense. Is it possible to refine the theorem by studying a barrier that is not a straight line?

The question is studied by Jaffuel [146]. Let $(a_i, i \geq 0)$ be a sequence of real numbers. We are interested in the probability that there exists an infinite ray (x_i) with $V(x_i) \leq a_i$ for all $i \geq 0$, which we call again the survival probability. What is the "critical barrier" for the survival probability to be positive?

In view of Theorem 5.34 in Sect. 5.7, one is tempted to think that the "critical barrier" should more or less look like a_i "$\approx$" $a_* i^{1/3}$ for large i, with $a_* := (\frac{3\pi^2\sigma^2}{2})^{1/3}$. It turns out that $i^{1/3}$ is indeed the correct order of magnitude for the critical barrier, but the constant a_* is incorrect.

Theorem 6.4 (Jaffuel [146]) *Assume $\psi(0) > 1$ and $\psi(1) = 0 = \psi'(1)$. If $\psi(1 + \delta) < \infty$ and $\mathbf{E}[(\#\Xi)^{1+\delta}] < \infty$ for some $\delta > 0$, then the probability*

$$\mathbf{P}(\exists \text{ infinite ray } (x_i) : V(x_i) \leq a\, i^{1/3}, \forall i \geq 0)$$

is positive if $a > a_c$ and vanishes if $a < a_c$, where $a_c := \frac{3}{2}(3\pi^2\sigma^2)^{1/3}$.

We observe that $a_c > a_*$. So for all $a \in (a_*, a_c)$, by Theorem 5.34 in Sect. 5.7, almost surely on the set of non-extinction, for all large n, there exist $(x_i, 0 \leq i \leq n)$ with $\varnothing =: x_0 < x_1 < \ldots < x_n$ and $|x_i| = i$, $0 \leq i \leq n$, such that $V(x_i) \leq a\, i^{1/3}$, $0 \leq i \leq n$, but no infinite ray satisfying the condition exists.

Theorem 6.4 does not tell us what happens if $a = a_c$.

Conjecture 6.5 Assume $\psi(0) > 1$ and $\psi(1) = 0 = \psi'(1)$. Under suitable integrability assumptions, we have

$$\mathbf{P}(\exists \text{ infinite ray } (x_i) : V(x_i) \leq a_c\, i^{1/3}, \forall i \geq 0) > 0,$$

where $a_c := \frac{3}{2}(3\pi^2\sigma^2)^{1/3}$.

6.2 The N-BRW

Starting from the 1990s, physicists have been interested in the slowdown phenomenon in the wave propagation of the F-KPP differential equation [72]. Instead of the standard F-KPP equation[1]

$$\frac{\partial u}{\partial t} = \frac{1}{2}\frac{\partial^2 u}{\partial x^2} + u(1-u),$$

with initial condition $u(0, x) = \mathbf{1}_{\{x<0\}}$, Brunet and Derrida [73] and Kessler et al. [153] introduced the cut-off version of the F-KPP equation:

$$\frac{\partial u}{\partial t} = \frac{1}{2}\frac{\partial^2 u}{\partial x^2} + u(1-u)\,\mathbf{1}_{\{u\geq\frac{1}{N}\}},$$

and discovered that the solution to the equation with cut-off has a wave speed that is slower than the standard speed by a difference of order $(\ln N)^{-2}$ when N is large.[2]

Later on, Brunet and Derrida [74] introduced a related F-KPP equation with white noise:

$$\frac{\partial u}{\partial t} = \frac{1}{2}\frac{\partial^2 u}{\partial x^2} + u(1-u) + \left(\frac{u(1-u)}{N}\right)^{1/2}\dot{W},$$

where $\dot{W}$ is the standard space-time white noise. [There is a duality between the noisy F-KPP equation and an appropriate reaction-diffusion system, see Doering et al. [97].] Once again, Brunet and Derrida found that the solution to the noisy F-KPP equation has a wave speed that is delayed, compared to the standard speed, by a quantity of order $(\ln N)^{-2}$ when N is large. This has been mathematically proved by Mueller et al. [199, 200].

On the other hand, the following so-called N-**BRW** was introduced by Brunet et al. [77–79]: in the branching random walk $(V(x))$, at each generation, only the N individuals having the smallest spatial values survive. The positions of the individuals in the resulting N-BRW are denoted by $(V^N(x))$. See Fig. 6.1 for an example with $N = 3$.

In order to avoid trivial discussions, we assume that there are no leaves in the branching random walks, i.e., $\#\varXi \geq 1$ with probability one. Since N is fixed, it is not hard to check that

$$v_N := \lim_{n\to\infty}\frac{1}{n}\max_{|x|=n} V^N(x) = \lim_{n\to\infty}\frac{1}{n}\min_{|x|=n} V^N(x),$$

---

[1]We have replaced u by $1 - u$ (thus considering the tail distribution, instead of the distribution function, of the maximum of branching Brownian motion) in the F-KPP equation (1.1) of Sect. 1.1.

[2]The notation is unfortunate, because N in this chapter has nothing to do with the random variable $\#\varXi$.

Fig. 6.1 An N-BRW with $N = 3$: first four generations

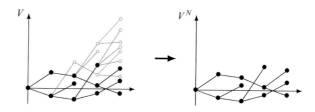

exists a.s., and is deterministic. Several predictions are made by these authors (see [77] in particular), for example, concerning the velocity v_N:

$$v_N = \frac{\pi^2\sigma^2}{2(\ln N)^2}\left(1 - \frac{(6 + o(1))\ln\ln N}{\ln N}\right), \quad N \to \infty, \tag{6.2}$$

where, as before, $\sigma^2 := \mathbf{E}(\sum_{|x|=1} V(x)^2 e^{-V(x)})$. [Of course, what is of particular interest in the conjectured precision $\frac{(6+o(1))\ln\ln N}{\ln N}$ is the universality of the main term.] All these predictions remain open, including a very interesting one concerning the genealogy of the particles in a suitable scale that would converge to the Bolthausen–Sznitman coalescent, though there is strong evidence that they are true in view of the recent progress made by Berestycki et al. [46].

However, the following result is remarkably proved by Bérard and Gouéré [40] by means of a rigorous argument. Recall that $\Xi := (\xi_1, \dots, \xi_{\#\Xi})$ is the point process governing the law of the branching random walk.

Theorem 6.6 (Bérard and Gouéré [40]) *Assume that $\#\Xi = 2$ and that ξ_1 are ξ_2 are i.i.d. If $\psi(1) = 0 = \psi'(1)$, and if $\psi(1 + \delta) < \infty$ and $\psi(-\delta) < \infty$ for some $\delta > 0$, then*

$$v_N = (1 + o(1))\frac{\pi^2\sigma^2}{2(\ln N)^2}, \quad N \to \infty.$$

The proof of Theorem 6.6 is technical, requiring several delicate couplings between the N-BRW and the usual branching random walk at an appropriate scale. We describe below a heuristic argument to see why v^N should behave asymptotically like $\frac{\pi^2\sigma^2}{2(\ln N)^2}$.

The basic idea is that the following two properties are "alike":

(a) A branching random walk, with an absorbing barrier of slope $\varepsilon > 0$ and starting with N particles at the origin, survives;
(b) An N-BRW moves at speed $\leq \varepsilon$.

In (a), the survival probability is $1 - (1 - p_{\text{surv}}(\varepsilon))^N$, where $p_{\text{surv}}(\varepsilon)$ is as in Theorem 6.1 (Sect. 6.1). This suggests that v_N would behave like $\varepsilon = \varepsilon(N)$ where ε

is defined by

$$p_{surv}(\varepsilon) \quad ``\approx" \quad \frac{1}{N}.$$

Solving the equation by means of Theorem 6.2 (Sect. 6.1), we obtain:

$$\varepsilon \sim \frac{\pi^2 \sigma^2}{2(\ln N)^2},$$

which gives Theorem 6.6.

Bérard and Gouéré in [40] succeed in making the heuristic argument rigorous. Unfortunately, the heuristic argument probably fails to lead to what is conjectured in (6.2). In other words, a deeper understanding of the N-BRW will be required for a proof of (6.2).

6.3   The L-BRW

Let $L > 0$. The following so-called L-**BRW** was introduced by Brunet et al. [77]: in the branching random walk $(V(x))$, at each generation, only the individuals whose spatial positions are within distance L to the minimal position are kept, while all others are removed from the system. We denote by $(V^{(L)}(x))$ the positions of the individuals in the L-BRW. Consider

$$v(L) := \lim_{n\to\infty} \frac{1}{n} \max_{|x|=n} V^{(L)}(x) = \lim_{n\to\infty} \frac{1}{n} \min_{|x|=n} V^{(L)}(x),$$

whenever it exists. It would be interesting to know how $v(L)$ behaves as $L \to \infty$. To the best of my knowledge, no non-trivial result is known in the literature.

Question 6.7   Study $v(L)$ as $L \to \infty$.

We can define the corresponding model for branching Brownian motion, called the L-BBM, and define the velocity of the system, $v_{BBM}(L)$. In [77], the authors argue that $v_{BBM}(L)$ should behave more or less like the velocity of the N-BBM (i.e., the analogue of the N-BRW for branching Brownian motion) if we take L to be of order $\ln N$. The following result is rigorously proved.

Theorem 6.8 (Pain [206]) *For the L-BBM, $v_{BBM}(L)$ exists for all $L > 0$, and*

$$v_{BBM}(L) = -2^{1/2} + (1 + o(1))\frac{\pi^2}{8^{1/2}L^2}, \quad L \to \infty.$$

The main term, $-2^{1/2}$, is simply the velocity of the minimal position in branching Brownian motion (see Sect. 1.1). So Theorem 6.8 is in agreement with Theorem 6.6 via the heuristics of [77].

6.4 Notes

Although the study of branching diffusions with absorption goes back to Sevast'yanov [220] and Watanabe [231], it is the work of Kesten [154] on branching Brownian motion with an absorbing barrier that is the most relevant to the topic in Sect. 6.1.

Theorem 6.1 is proved by Biggins et al. [59] under stronger assumptions, excluding the case $\varepsilon = 0$.

Theorem 6.2 is proved in [115]; see also [41] for a different proof, which moreover gives some additional precision on the $o(1)$ expression in (6.1). For branching Brownian motion, (much) more is known, see Aïdékon and Harris [12], Berestycki et al. [45, 46], Harris et al. [128].

A related problem of survival probability for the branching random walk with absorption concerns the critical slope $\varepsilon = 0$ and the situation that the system survives in the first n steps. This is studied in depth in Aïdékon and Jaffuel [13]. The corresponding problem for branching Brownian motion with absorption is investigated in the pioneering work of Kesten [154], and improved by Harris and Harris [124]. For links with the one-sided F-KPP equation, see Harris et al. [127]. For physics literature, see Derrida and Simon [93, 222], where many interesting predictions are made.

The analogue of Theorem 6.3 for branching Brownian motion is proved by Maillard [185], who is moreover able to obtain an accurate evaluation for the density of $\#\mathscr{Z}$ at infinity.

The analogue of Theorem 6.4 for branching Brownian motion is proved by Roberts [216], with remarkable precision. In particular, the analogue of Conjecture 6.5 for branching Brownian motion is proved in [216].

In addition of Theorem 6.6 (Sect. 6.2) proved by Bérard and Gouéré [40], other rigorous results concerning the N-BRW (or the analogue for branching Brownian motion) are obtained by Durrett and Remenik [104], Maillard [186], Bérard and Maillard [42], Mallein [191, 192], and by Berestycki and Zhao [44] in higher dimensions. In particular, the weak convergence of the empirical measure of the N-BBM (the analogue of the N-BRW for branching Brownian motion) is proved in [104].

For both N-BBM and L-BBM, deviation properties are studied in [92].

Chapter 7
Biased Random Walks on Galton–Watson Trees

This chapter is a brief presentation of the randomly biased random walk on trees in its slow regime. The model has been introduced by Lyons and Pemantle [174], as an extension of Lyons's deterministically biased random walk on trees [171, 172].

7.1 A Simple Example

Before introducing the general model, let us start with a simple example.

Example 7.1 Consider a rooted regular binary tree, and add a parent $\overleftarrow{\varnothing}$ to the root $\varnothing$.[1] The resulting tree is a planted tree (in the sense of [100]). We give a random colour to each of the vertices of the tree; a vertex is coloured red with probability p_{red}, and blue with probability p_{blue}, with $p_{\text{red}} > 0$ and $p_{\text{blue}} > 0$ such that $p_{\text{red}} + p_{\text{blue}} = 1$.

A random walker performs a discrete-time random walk on the tree, starting from the root $\varnothing$. At each step, the walk stays at a vertex for a unit of time, then moves to one of the neighbours (either the parent, or one of the two children). The transition probabilities are $a_{\text{red}}^{\uparrow}$ (moving to the parent), $a_{\text{red}}^{(1)}$ and $a_{\text{red}}^{(2)}$ (moving to either of the children) if the site where the walker stays currently is red, or $a_{\text{blue}}^{\uparrow}$, $a_{\text{blue}}^{(1)}$ and $a_{\text{blue}}^{(2)}$ if the site is blue. We assume that $a_{\text{red}}^{\uparrow}$, $a_{\text{red}}^{(1)}$, $a_{\text{red}}^{(2)}$, $a_{\text{blue}}^{\uparrow}$, $a_{\text{blue}}^{(1)}$ and $a_{\text{blue}}^{(2)}$ are positive numbers such that

$$a_{\text{red}}^{\uparrow} + a_{\text{red}}^{(1)} + a_{\text{red}}^{(2)} = 1 = a_{\text{blue}}^{\uparrow} + a_{\text{blue}}^{(1)} + a_{\text{blue}}^{(2)}.$$

---

[1] The root $\varnothing$ is a vertex of the tree, but $\overleftarrow{\varnothing}$ is not considered as a vertex of the tree.

© Springer International Publishing Switzerland 2015
Z. Shi, *Branching Random Walks*, Lecture Notes in Mathematics 2151,
DOI 10.1007/978-3-319-25372-5_7

The parent $\overleftarrow{\varnothing}$ of the root is reflecting: Each time the walk is at $\overleftarrow{\varnothing}$, it automatically comes back to $\varnothing$ in the next step. $\qquad\square$

The usual questions arise naturally: Is the random walk recurrent or transient? What can be said about its position after n steps? What is the maximal displacement in the first n steps?

7.2 The Slow Movement

Let $\mathbb{T}$ be a supercritical Galton–Watson tree; we add a parent $\overleftarrow{\varnothing}$ to the root $\varnothing$. For any $x \in \mathbb{T}$, let $\overleftarrow{x}$ denote the parent of x (recalling that $\overleftarrow{\varnothing}$ is not considered as a vertex of $\mathbb{T}$), and $x^{(1)}, \ldots, x^{(N(x))}$ the children of x. Let $\omega := (\omega(x), x \in \mathbb{T})$ be a family of i.i.d. random vectors, with $\omega(x) = (\omega(x, y), y \in \{\overleftarrow{x}\} \cup \{x^{(1)}, \ldots, x^{(N(x))}\})$. We assume that with probability one, $\omega(\varnothing, y) > 0$ for $y \in \{\overleftarrow{\varnothing}\} \cup \{\varnothing^{(1)}, \ldots, \varnothing^{(N(\varnothing))}\}$, and that $\omega(\varnothing, \overleftarrow{\varnothing}) + \sum_{i=1}^{N(x)} \omega(\varnothing, \varnothing^{(i)}) = 1$. In Example 7.1, $\omega(\varnothing)$ (or any $\omega(x)$, for $x \in \mathbb{T}$) is a three-dimensional random vector and takes two possible values $(a_{red}^{\uparrow}, a_{red}^{(1)}, a_{red}^{(2)})$ and $(a_{blue}^{\uparrow}, a_{blue}^{(1)}, a_{blue}^{(2)})$ with probability p_{red} and p_{blue}, respectively.

For each given ω (which, in Example 7.1, means that all the colours are known), let $(X_n, n \geq 0)$ be a Markov chain with $X_0 = \varnothing$ and with transition probabilities

$$P_\omega(X_{n+1} = \overleftarrow{x} \mid X_n = x) = \omega(x, \overleftarrow{x}),$$
$$P_\omega(X_{n+1} = x^{(i)} \mid X_n = x) = \omega(x, x^{(i)}), \quad 1 \leq i \leq N(x),$$

and $P_\omega(X_{n+1} = y \mid X_n = x) = 0$ if $y \notin \{\overleftarrow{x}\} \cup \{x^{(1)}, \ldots, x^{(N(x))}\}$.

We use $\mathbf{P}$ to denote the probability with respect to the environment, and $\mathbb{P} := \mathbf{P} \otimes P_\omega$ the annealed probability, i.e., $\mathbb{P}(\cdot) := \int P_\omega(\cdot) \mathbf{P}(d\omega)$.

A convenient way to study the effect of ω on the behaviour of (X_n) is via the following process $(V(x), x \in \mathbb{T})$ defined by $V(\varnothing) := 0$ and

$$V(x) = \sum_{i=0}^{|x|-1} \ln \frac{\omega(x_i, x_{i-1})}{\omega(x_i, x_{i+1})}, \quad x \in \mathbb{T}\backslash\{\varnothing\}, \quad (x_{-1} := \overleftarrow{\varnothing})$$

where x_i denotes, as before, the ancestor of x in the i-th generation (for $0 \leq i \leq |x|$). It is immediately seen that $(V(x), x \in \mathbb{T})$ is a branching random walk studied in the previous chapters! Let $\varXi := (\xi_1, \cdots, \xi_N)$ be the point process governing the law of the branching random walk, i.e., $(V(x), |x| = 1)$ is distributed as $\varXi$.

Example 7.2 If $\varXi = (\xi_1, \ldots, \xi_N) = (\ln \lambda, \ldots, \ln \lambda)$, where $\lambda > 0$ is a fixed parameter, the resulting process $(X_n, n \geq 0)$ is Lyons's λ-biased random walk on Galton–Watson trees [171, 172]. $\qquad\square$

Let, as before, $\psi(t) := \ln \mathbf{E}(\sum_{|x|=1} e^{-V(x)})$, $t \in \mathbb{R}$. Throughout the chapter, we assume $\psi(0) > 0$, and $\psi(1) = 0 = \psi'(1)$.

Theorem 7.3 (Lyons and Pemantle [174]) *Assume $\psi(0) > 0$, and $\psi(1) = 0 = \psi'(1)$. The biased random walk $(X_n, n \ge 0)$ is $\mathbb{P}$-almost surely recurrent.*

Proof There is nothing to prove if the Galton–Watson tree $\mathbb{T}$ is finite. So let us work on the set of non-extinction of $\mathbb{T}$. Write $T_x := \inf\{i \ge 0 : X_i = x\}$, the first hitting time of vertex x, and $T_\varnothing^+ := \inf\{i \ge 1 : X_i = \varnothing\}$, the first *return* time to the root, with $\inf \varnothing := \infty$.

Let $x \in \mathbb{T}$ with $|x| = n \ge 1$. For any $0 \le k \le n$, write $a_k := P_\omega\{T_x < T_\varnothing \mid X_0 = x_k\}$. Then $a_0 = 0$, $a_n = 1$, and for $1 \le k < n$,

$$a_k = \frac{\omega(x_k, x_{k+1})}{\omega(x_k, x_{k+1}) + \omega(x_k, x_{k-1})} a_{k+1} + \frac{\omega(x_k, x_{k-1})}{\omega(x_k, x_{k+1}) + \omega(x_k, x_{k-1})} a_{k-1}.$$

Solving the system of linear equations leads to

$$P_\omega\{T_x < T_\varnothing \mid X_0 = x_1\} = a_1 = \frac{e^{V(x_1)}}{\sum_{i=1}^n e^{V(x_i)}}.$$

In particular,[2]

$$P_\omega\{T_x < T_\varnothing^+\} = \omega(\varnothing, x_1) P_\omega\{T_x < T_\varnothing \mid X_0 = x_1\} = \frac{\omega(\varnothing, \overleftarrow{\varnothing})}{\sum_{i=1}^n e^{V(x_i)}}.$$

Let $\tau_n := \inf\{i \ge 0 : |X_i| = n\}$. Then for $n \ge 1$,

$$P_\omega\{\tau_n < T_\varnothing^+\} \le \sum_{x \in \mathbb{T} : |x| = n} P_\omega\{T_x < T_\varnothing^+\} \le \omega(\varnothing, \overleftarrow{\varnothing}) \sum_{x \in \mathbb{T} : |x| = n} \frac{1}{\sum_{i=1}^n e^{V(x_i)}},$$

$$\tag{7.1}$$

which yields $P_\omega\{\tau_n < T_\varnothing^+\} \le \sum_{x \in \mathbb{T} : |x| = n} e^{-V(x)}$. Consequently, $P_\omega\{\tau_n < T_\varnothing^+\} \to 0$ (as $n \to \infty$) almost surely on the set of non-extinction (by the Biggins martingale convergence theorem, Theorem 3.2 in Sect. 3.2). The recurrence follows immediately. □

If $\psi(0) > 0$ and $\psi(1) = 0 = \psi'(1)$, the biased random walk is null recurrent (Faraud [110]; under some additional integrability conditions). Theorem 7.4 below tells us that the biased random walk is very slow.

Let $(\mathfrak{m}(s), s \in [0, 1])$ be a standard Brownian meander under $\mathbf{P}$, and let $\overline{\mathfrak{m}}(s) := \sup_{u \in [0, s]} \mathfrak{m}(u)$. Recall that the standard Brownian meander can be realized

---

[2]This simple formula tells us that V plays the role of **potential**: The higher the potential value is on the path $\{x_1, \ldots, x_n\}$, the harder it is for the biased random walk to reach x.

as $\mathrm{m}(s) := \frac{|B(\mathfrak{g}+s(1-\mathfrak{g}))|}{(1-\mathfrak{g})^{1/2}}$, $s \in [0, 1]$, where $(B(t), t \in [0, 1])$ is a standard Brownian motion, with $\mathfrak{g} := \sup\{t \le 1 : B(t) = 0\}$.

Theorem 7.4 (Hu and Shi [139]) *Assume $\psi(0) > 0$ and $\psi(1) = 0 = \psi'(1)$. If $\psi(1 + \delta) < \infty$, $\psi(-\delta) < \infty$ and $\mathbf{E}[(\#\varXi)^{1+\delta}] < \infty$ for some $\delta > 0$, then for all $u \ge 0$,*

$$\lim_{n\to\infty} \mathbb{P}\left(\frac{\sigma^2 |X_n|}{(\ln n)^2} \le u \,\middle|\, \text{non-extinction}\right) = \int_0^u \frac{1}{(2\pi r)^{1/2}} \mathbf{P}\left(\eta \le \frac{1}{r^{1/2}}\right) dr,$$

where $\sigma^2 := \mathbf{E}(\sum_{|x|=1} V(x)^2 e^{-V(x)}) \in (0, \infty)$ as before, and $\eta := \sup_{s\in[0, 1]}[\overline{\mathrm{m}}(s) - \mathrm{m}(s)]$.

We mention that $\int_0^\infty \frac{1}{(2\pi r)^{1/2}} \mathbf{P}(\eta \le \frac{1}{r^{1/2}}) dr = 1$ because $\mathbf{E}(\frac{1}{\eta}) = (\frac{\pi}{2})^{1/2}$, see [140]. Very recently, Pitman [212] obtains an analytical expression for the distribution function of η, by means of a result of Biane and Yor [48]; it is proved that η has the Kolmogorov–Smirnov distribution: for $x > 0$,

$$\mathbf{P}(\eta \le x) = \sum_{k=-\infty}^{\infty} (-1)^k e^{-2k^2 x^2} = \frac{(2\pi)^{1/2}}{x} \sum_{j=0}^{\infty} \exp\left(-\frac{(2j + 1)^2 \pi^2}{8x^2}\right).$$

Since the biased random walk on trees can be viewed as a random walk in random environment on trees, Theorem 7.4 is the analogue on trees of Sinai [223]'s result for one-dimensional random walk in random environment. We mention that like in Sinai's case, there is a localization result for the biased random walk on trees; see [139].

7.3 The Maximal Displacement

Let $\varXi$ denote as before a point process having the law of $(V(x), |x| = 1)$. Assume that $\psi(0) > 1$, $\psi(1) = 0 = \psi'(1)$, and that $\#\varXi \ge 1$ a.s. Let as before $\tau_n := \inf\{i \ge 0 : |X_i| = n\}$. By (7.1),

$$P_\omega\{\tau_n < T_\varnothing^+\} \le \sum_{x\in\mathbb{T}: |x|=n} e^{-\overline{V}(x)},$$

with $\overline{V}(x) := \max_{1\le i\le |x|} V(x_i)$. By Theorem 5.34 in Sect. 5.7 (under its assumptions), we have, with $a_* := (\frac{3\pi^2\sigma^2}{2})^{1/3}$ and $\sigma^2 := \mathbf{E}(\sum_{|x|=1} V(x)^2 e^{-V(x)})$,

$$\liminf_{n\to\infty} \frac{1}{n^{1/3}} \ln P_\omega\{\tau_n < T_\varnothing^+\} \ge -a_*, \quad \mathbf{P}\text{-a.s.}$$

Let $L_k := \sum_{i=1}^{k} \mathbf{1}_{\{X_i = \varnothing\}}$, which stands for the number of visits at the root $\varnothing$ in the first k steps. Then for any $j \geq 1$ and all $\varepsilon > 0$,

$$P_\omega\{L_{\tau_n} \geq j\} = [P_\omega\{\tau_n > T_\varnothing^+\}]^j \leq \left[1 - e^{-(1+\varepsilon)a_* n^{1/3}}\right]^j \leq \exp\left(-je^{-(1+\varepsilon)a_* n^{1/3}}\right),$$

P-almost surely for all sufficiently large n (say $n \geq n_0(\omega)$; $n_0(\omega)$ does not depend on j). Taking $j := \lfloor e^{(1+2\varepsilon)a_* n^{1/3}} \rfloor$, we see that

$$\sum_n P_\omega\{L_{\tau_n} \geq \lfloor e^{(1+2\varepsilon)a_* n^{1/3}} \rfloor\} < \infty, \quad \textbf{P}\text{-a.s.}$$

This implies, by the Borel–Cantelli lemma, that

$$\limsup_{n \to \infty} \frac{\ln L_{\tau_n}}{n^{1/3}} \leq a_*, \quad \textbf{P}\text{-a.s.}$$

It is known, and not hard, to check that

$$\lim_{k \to \infty} \frac{\ln L_k}{\ln k} = 1, \quad \textbf{P}\text{-a.s.},$$

which yields that

$$\limsup_{n \to \infty} \frac{\ln \tau_n}{n^{1/3}} \leq a_*, \quad \textbf{P}\text{-a.s.}$$

Note that for all n and j, $\{\tau_n \leq k\} = \{\max_{1 \leq i \leq k} |X_i| \geq n\}$. This implies that

$$\liminf_{n \to \infty} \frac{1}{(\ln n)^3} \max_{1 \leq i \leq n} |X_i| \geq \frac{1}{a_*^3} = \frac{2}{3\pi^2 \sigma^2}, \quad \textbf{P}\text{-a.s.}$$

It turns out that $(\ln n)^3$ is the correct order of magnitude for $\max_{1 \leq i \leq n} |X_i|$, but the constant $\frac{2}{3\pi^2 \sigma^2}$ is not optimal.

Theorem 7.5 (Faraud et al. [111]) *Assume $\psi(0) > 0$ and $\psi(1) = 0 = \psi'(1)$. If $\psi(1 + \delta) < \infty$, $\psi(-\delta) < \infty$ and $\mathbf{E}[(\#\Xi)^{1+\delta}] < \infty$ for some $\delta > 0$, then $\mathbb{P}$-almost surely on the set of non-extinction,*

$$\lim_{n \to \infty} \frac{1}{(\ln n)^3} \max_{1 \leq i \leq n} |X_i| = \frac{8}{3\pi^2 \sigma^2}.$$

Theorems 7.4 and 7.5 together reveal a multifractal structure in the sample path of the biased random walk. Loosely speaking, the biased random walk typically stays at a distance of order $(\ln n)^2$ to the root after n steps, but at some exceptional times during the first n steps, the biased random walk makes a displacement of order $(\ln n)^3$ to the root. It would be interesting to quantify this phenomenon and to connect, in some sense, the two theorems.

Let $\mathfrak{G}_n := \sup\{i \le n : X_i = \varnothing\}$, the last passage time at the root before n.

Conjecture 7.6 Assume $\psi(0) > 0$ and $\psi(1) = 0 = \psi'(1)$. Under suitable integrability assumptions, $\mathbb{P}$-almost surely on the set of non-extinction,

$$\limsup_{n\to\infty} \frac{1}{(\ln n)^3} \max_{\mathfrak{G}_n \le i \le n} |X_i| > 0.$$

7.4   Favourite Sites

For any vertex $x \in \mathbb{T}$, let

$$L_n(x) := \sum_{i=1}^{n} \mathbf{1}_{\{X_i = x\}}, \quad n \ge 1,$$

which is the site local time at position x of the biased random walk. Consider, for any $n \ge 1$, the set of the favourite sites (or: most visited sites) at time n:

$$\mathscr{A}_n := \left\{ x \in \mathbb{T} : L_n(x) = \max_{y \in \mathbb{T}} L_n(y) \right\}.$$

The study of favourite sites was initiated by Erdős and Révész [105] for the symmetric simple random walk on $\mathbb{Z}$ (see a list of ten open problems presented in Chap. 11 of Révész [214]) who conjectured for the latter process that the family of favourite sites is tight, and that $\#\mathscr{A}_n \le 2$ almost surely for all sufficiently large n. The second conjecture received a partial answer from Tóth [227], and is believed to be true. The first conjecture was disproved by Bass and Griffin [35], who proved that $\inf\{|x|, \ x \in \mathscr{A}_n\} \to \infty$ almost surely for the symmetric simple random walk on $\mathbb{Z}$; later, it was proved to be also the case [134] for Sinai's one-dimensional random walk in random environment.

Theorem 7.7 (Hu and Shi [138]) *Assume $\psi(0) > 0$ and $\psi(1) = 0 = \psi'(1)$. If $\psi(1 + \delta) < \infty$, $\psi(-\delta) < \infty$ and $\mathbb{E}[(\#\Xi)^{1+\delta}] < \infty$ for some $\delta > 0$, then there exists a finite non-empty set $\mathscr{U}$, depending only on the environment, such that*

$$\lim_{n\to\infty} \mathbb{P}(\mathscr{A}_n \subset \mathscr{U} \mid non\text{-}extinction) = 1.$$

In particular, the family of most visited sites is tight under $\mathbb{P}$.

Theorem 7.7 tells us that as far as the tightness question for favourite sites is concerned, the biased random walk on trees behaves very differently from the recurrent nearest-neighbour random walk on $\mathbb{Z}$ (whether the environment is random or deterministic). Il also gives a non-trivial example of null recurrent Markov chain whose favourite sites are tight.

We close this brief chapter with a question.

Question 7.8 Assume $\psi(0) > 0$ and $\psi(1) = 0 = \psi'(1)$. Under suitable integrability assumptions, is it true that $\limsup_{n\to\infty} \sup_{x\in\mathscr{A}_n} |x| < \infty$ $\mathbb{P}$-almost surely?

7.5 Notes

This chapter has shown only a tiny branch (on which I sit) of a big tree in the forest of probability on trees. For a better picture of the forest, the book [175] and the lectures notes [208] are ideal references.

Theorem 7.3 in Sect. 7.2 is a special case of a general result of Lyons and Pemantle [174], who give a recurrence/transience criterion for random walks on general trees (not only Galton–Watson trees). For a proof using Mandelbrot's multiplicative cascades, see Menshikov and Petritis [195].

The biased random walk on trees is often viewed as a random walk in random environment on trees. The elementary proof of Theorem 7.3 presented here is indeed essentially borrowed from Solomon [224] for one-dimensional random walks in random environment. For more elaborated techniques and a general account of random walks in random environment, see the lecture notes of Zeitouni [234].

Related to discussions in Sect. 7.4, local time of generations (instead of site local times as in Sect. 7.4) for the randomly biased random walk is investigated by Andreoletti and Debs in [24, 25], and site local time in the sub-diffusive regime by Hu [133].

There is huge literature on random walks on trees. Let me attempt to make a rather incomplete list of works: they are here either because they are connected to the material presented in this chapter, or because they concern velocity and related questions.

I only include results for biased random walks on supercritical Galton–Watson trees, referred to simply as "biased random walks" below.

It is already pointed out that a general recurrence/transience criterion is available from Lyons and Pemantle [174].

(a) In the **transient** case, a general result of Gross [118] shows the existence of the velocity. A natural question to answer is what can be said about the velocity, and whether or not the velocity is positive.

(a1) For the λ-**biased random walk** (on trees), transience means $\lambda < \mathbf{E}(\#\varXi)$, the answer of positivity of the velocity depends on whether or not there are leaves on trees.

When $\#\varXi \geq 1$ $\mathbf{P}$-a.s., the tree is leafless, the velocity of the biased random walk is positive [178]. For simple random walk (i.e., $\lambda = 1$), the value of the velocity is explicitly known [177]. For other values of λ, there is a simple and nice upper bound [86, 230]. More recently, Aïdékon [10] deduces an analytical expression for the velocity. A famous conjecture [178] is that the velocity is non-increasing in $\lambda \in [0, \mathbf{E}(\#\varXi))$. This monotonicity is established by Ben Arous et al. [38] for λ in the neighbourhood of $\mathbf{E}(\#\varXi)$, by Ben Arous et al. [39] for λ in the neighbourhood of 0, and by Aïdékon [9] for $\lambda \in [0, \frac{1}{2}]$ using his analytical expression obtained in [10]. Another interesting question concerns the smoothness of the velocity as a function of λ: the function is proved in [178] to be continuous on $[0, 1)$, but it is not known if it is continuous on $[1, \mathbf{E}(\#\varXi))$. In the sub-ballistic case (i.e., with zero velocity), a quenched invariance principle is proved by Peres and Zeitouni [209].

When $\mathbf{P}(\#\varXi = 0) > 0$, the tree has leaves. It is proved in [178] that the velocity is positive if and only if $\lambda > f'(q)$, where f is the moment generating function of the reproduction law of the Galton–Watson tree as in Sect. 2.1, and q is the extinction probability. In the sub-ballistic case, the escape rate is studied by Ben Arous et al. [37], who also prove a cyclic phenomenon (tightness but no weak convergence).

Regardless of existence of leaves, an annealed invariance principle is proved by Piau [211].

(a2) We now turn to the **randomly biased random walk**.

When $\#\varXi \geq 1$ **P**-a.s., Aïdékon gives in [5] a criterion for the positivity of the velocity, and in [6] a simple upper bound for the velocity in case it is positive, which coincides with the aforementioned upper bound of Chen [86] and Virág [230] when there is no randomness in the bias. In the sub-ballistic case, the escape rate is also studied in [5]. Deviation properties are studied in [6].

Assume now $\mathbf{P}(\#\varXi = 0) > 0$. In the sub-ballistic case, weak convergence is proved by Hammond [121] (contrasting with the λ-biased random walk).

(b) For **recurrent** randomly biased random walk, the maximal displacement is studied in [135] (the sub-diffusive regime) and in [136] (the slow regime studied in this chapter). Faraud [110] establishes a quenched invariance principle, extending the aforementioned result of Peres and Zeitouni [209] for the λ-biased random walk in the very delicate null recurrent case.

Deviations properties, in both quenched and annealed settings, are studied by Dembo et al. [91] for the λ-biased random walk (leafless, recurrent or transient), and by Aïdékon [6] for the randomly biased random walk (leafless, transient).

For a general account of the biased random walk on trees, see (again) [175].

Appendix A
Sums of i.i.d. Random Variables

Let $(S_n - S_{n-1}, n \geq 1)$ be a sequence of independent and identically real-valued random variables such that $\mathbf{E}(S_1 - S_0) = 0$ and that $\mathbf{P}(S_1 - S_0 \neq 0) > 0$. We assume $S_0 = 0$, $\mathbf{P}$-a.s.

A.1   The Renewal Function

The material in this paragraph is well-known; see, for example, Feller [112].

Define the function $R : [0, \infty) \to (0, \infty)$ by $R(0) := 1$ and

$$R(u) := \mathbf{E}\left\{ \sum_{j=0}^{\tau^+ - 1} \mathbf{1}_{\{S_j \geq -u\}} \right\}, \quad u > 0, \tag{A.1}$$

where $\tau^+ := \inf\{k \geq 1 : S_k \geq 0\}$ (which is well-defined almost surely, since $\mathbf{E}(S_1) = 0$).

The function R is the **renewal function** associated with (S_n), because it can be written as

$$R(u) = \sum_{k=0}^{\infty} \mathbf{P}\{H_k^- \geq -u\} = \sum_{k=0}^{\infty} \mathbf{P}\{|H_k^-| \leq u\}, \quad u \geq 0, \tag{A.2}$$

where $H_0^- > H_1^- > H_2^- > \ldots$ are the strictly descending ladder heights of (S_n), i.e.,

$$H_k^- := S_{\theta_k^-}, \quad k \geq 0, \tag{A.3}$$

© Springer International Publishing Switzerland 2015
Z. Shi, *Branching Random Walks*, Lecture Notes in Mathematics 2151,
DOI 10.1007/978-3-319-25372-5

with $\theta_0^- := 0$ and $\theta_k^- := \inf\{i > \theta_{k-1}^- : S_i < \min_{0 \le j \le \theta_{k-1}^-} S_j\}$ for $k \ge 1$. By the renewal theorem,

$$\frac{R(u)}{u} \to c_{\text{ren}} \in [0, \infty), \quad u \to \infty, \tag{A.4}$$

where $c_{\text{ren}} := \frac{1}{\mathbf{E}(|H_1^-|)}$. Moreover [99], if $\mathbf{E}\{[(S_1)_-]^2\} < \infty$ (where $u_- := \max\{-u, 0\}$ for $u \in \mathbb{R}$), then

$$0 < c_{\text{ren}} < \infty. \tag{A.5}$$

A.2 Random Walks to Stay Above a Barrier

Throughout this part, we assume, moreover, that $\mathbf{E}(S_1^2) \in (0, \infty)$.

We list a few elementary probability estimates for (S_n) to stay above a barrier; these estimates are useful in various places of the notes. They are collected or adapted from the appendices of [8, 14, 15], but many of them can probably be found elsewhere in the literature in one form or another. For a survey on general ballot-type theorems, see Addario-Berry and Reed [2].

Sometimes we work under $\mathbf{P}_v$ (for $v \in \mathbb{R}$), meaning that $\mathbf{P}_v(S_0 = v) = 1$ (so $\mathbf{P}_0 = \mathbf{P}$); the corresponding expectation is denoted by $\mathbf{E}_v$.

Let us write

$$\underline{S}_n := \min_{0 \le i \le n} S_i, \quad n \ge 0.$$

By Stone's local limit theorem, there exist constants $c_{55} > 0$ and $c_{56} > 0$ such that

$$\sup_{r \in \mathbb{R}} \mathbf{P}\{r \le S_n \le r + h\} \le c_{55} \frac{h}{n^{1/2}}, \quad \forall n \ge 1, \ \forall h \ge c_{56}. \tag{A.6}$$

It is known [159] that there exist positive constants C_+ and C_- such that for $a \ge 0$ and $n \to \infty$,

$$\mathbf{P}\left\{\min_{0 \le i \le n} S_i \ge -a\right\} \sim \frac{C_+ R(a)}{n^{1/2}}, \quad \mathbf{P}\left\{\max_{0 \le i \le n} S_i \le -a\right\} \sim \frac{C_- R_-(a)}{n^{1/2}}, \tag{A.7}$$

where R_- is the renewal function associated with $(-S_n)$. Furthermore, it is possible to have a bound in (A.7) which is valid uniformly in a [159]:

$$\limsup_{n \to \infty} n^{1/2} \sup_{a \ge 0} \frac{1}{a+1} \mathbf{P}\left\{\underline{S}_n \ge -a\right\} < \infty. \tag{A.8}$$

With the notation $x \wedge y := \min\{x, y\}$, we claim that there exists $c_{57} > 0$ such that for $a \geq 0$, $b \geq -a$ and $n \geq 1$,

$$\mathbf{P}\Big\{b \leq S_n \leq b + c_{56}, \ \underline{S}_n \geq -a\Big\} \leq c_{57} \frac{[(a+1) \wedge n^{1/2}] \, [(b+a+1) \wedge n^{1/2}]}{n^{3/2}},$$
(A.9)

where $c_{56} > 0$ is the constant in (A.6). We only need to prove it for all sufficiently large n; for notational simplification, we treat $\frac{n}{3}$ as an integer. By the Markov property at time $\frac{n}{3}$,

$$\mathbf{P}\Big\{b \leq S_n \leq b + c_{56}, \ \underline{S}_n \geq -a\Big\}$$

$$\leq \mathbf{P}\Big\{\underline{S}_{\frac{n}{3}} \geq -a\Big\} \sup_{x \geq -a} \mathbf{P}\Big\{b - x \leq S_{\frac{2n}{3}} \leq b - x + c_{56}, \ \underline{S}_{\frac{2n}{3}} \geq -a - x\Big\}.$$

By (A.8), $\mathbf{P}\{\underline{S}_{\frac{n}{3}} \geq -a\} \leq c_{58} \frac{(a+1) \wedge n^{1/2}}{n^{1/2}}$. It remains to check that

$$\sup_{x \geq -a} \mathbf{P}\Big\{b - x \leq S_{\frac{2n}{3}} \leq b - x + c_{56}, \ \underline{S}_{\frac{2n}{3}} \geq -a - x\Big\} \leq c_{62} \frac{(b+a+1) \wedge n^{1/2}}{n}.$$

Let $\widetilde{S}_j := S_{\frac{2n}{3} - j} - S_{\frac{2n}{3}}$. Then $\mathbf{P}\{b - x \leq S_{\frac{2n}{3}} \leq b - x + c_{56}, \ \underline{S}_{\frac{2n}{3}} \geq -a - x\} \leq \mathbf{P}\{-b + x - c_{56} \leq \widetilde{S}_{\frac{2n}{3}} \leq -b + x, \ \min_{1 \leq i \leq \frac{2n}{3}} \widetilde{S}_i \geq -a - b - c_{56}\}$. By the Markov property, this leads to: for $x \geq -a$,

$$\mathbf{P}\Big\{b - x \leq S_{\frac{2n}{3}} \leq b - x + c_{56}, \ \underline{S}_{\frac{2n}{3}} \geq -a - x\Big\}$$

$$\leq \mathbf{P}\Big\{\min_{1 \leq i \leq \frac{n}{3}} \widetilde{S}_i \geq -a - b - c_{56}\Big\} \sup_{y \in \mathbb{R}} \mathbf{P}\Big\{-b + x - c_{56} - y \leq \widetilde{S}_{\frac{n}{3}} \leq -b + x - y\Big\}.$$

The first probability expression on the right-hand side is bounded by $c_{60} \frac{(b+a+1) \wedge n^{1/2}}{n^{1/2}}$ (by (A.8)), whereas the second by $\frac{c_{61}}{n^{1/2}}$ (by (A.6)). So (A.9) is proved.

Lemma A.1 *There exists $c_{62} > 0$ such that for $a \geq 0$, $b \geq -a$ and $n \geq 1$,*

$$\mathbf{P}\Big\{\underline{S}_n \geq -a, \ S_n \leq b\Big\} \leq c_{62} \frac{[(a+1) \wedge n^{1/2}] \, [(b+a+1)^2 \wedge n]}{n^{3/2}}.$$

Proof It is a straightforward consequence of (A.9). □

Lemma A.2 *There exists $c_{63} > 0$ such that for $u > 0$, $a \geq 0$, $b \geq 0$ and $n \geq 1$,*

$$\mathbf{P}\Big\{\underline{S}_n \geq -a, \ b - a \leq S_n \leq b - a + u\Big\} \leq c_{63} \frac{(u+1)(a+1)(b+u+1)}{n^{3/2}}.$$

Proof The inequality follows immediately from Lemma A.1 if $u = c_{56}$, and thus holds also for $u < c_{56}$, whereas the case $u > c_{56}$ boils down to the case $u = c_{56}$ by splitting $[b - a, b - a + u]$ into intervals of lengths $\leq c_{56}$, the number of these intervals being less than $(\frac{u}{c_{56}} + 1)$.

[When $a = 0$, both Lemmas A.1 and A.2 boil down to a special case of Lemma 20 of Vatutin and Wachtel [229].]   □

Lemma A.3 *There exists $c_{64} > 0$ such that for $a \geq 0$,*

$$\sup_{n \geq 1} \mathbf{E}\left[|S_n| \, \mathbf{1}_{\{\underline{S}_n \geq -a\}} \right] \leq c_{64} \, (a + 1).$$

Proof We need to check that for some $c_{65} > 0$, $\mathbf{E}[S_n \, \mathbf{1}_{\{\underline{S}_n \geq -a\}}] \leq c_{65} \, (a+1)$, $\forall a \geq 0$, $\forall n \geq 1$.

Let $\tau_a^- := \inf\{i \geq 1 : S_i < -a\}$. Then $\mathbf{E}[S_n \, \mathbf{1}_{\{\underline{S}_n \geq -a\}}] = -\mathbf{E}[S_n \, \mathbf{1}_{\{\tau_a^- \leq n\}}]$, which, by the optional sampling theorem, is equal to $\mathbf{E}[(-S_{\tau_a^-}) \, \mathbf{1}_{\{\tau_a^- \leq n\}}]$. As a consequence, we have $\sup_{n \geq 1} \mathbf{E}[S_n \, \mathbf{1}_{\{\underline{S}_n \geq -a\}}] = \mathbf{E}[(-S_{\tau_a^-})]$.

It remains to check that $\mathbf{E}[(-S_{\tau_a^-}) - a] \leq c_{66} \, (a + 1)$ for some $c_{66} > 0$ and all $a \geq 0$. [In fact, assuming $\mathbf{E}(|S_1|^3) < \infty$, it is even true that $\sup_{a \geq 0} \mathbf{E}[(-S_{\tau_a^-}) - a] < \infty$; see Mogulskii [196].] By a well-known trick [163] using the sequence of strictly descending ladder heights, it boils down to proving that $\mathbf{E}[(-\widetilde{S}_{\widetilde{\tau}_a^-}) - a] \leq c_{67} \, (a+1)$ for some $c_{67} > 0$ and all $a \geq 0$, where $\widetilde{S}_1, \widetilde{S}_2 - \widetilde{S}_1, \widetilde{S}_3 - \widetilde{S}_2, \ldots$ are i.i.d. *negative* random variables with $\mathbf{E}(\widetilde{S}_1) > -\infty$, and $\widetilde{\tau}_a^- := \inf\{i \geq 1 : \widetilde{S}_i < -a\}$. This, however, is a special case of (2.6) of Borovkov and Foss [62].   □

Lemma A.4 *Let $0 < \lambda < 1$. There exists $c_{68} > 0$ such that for $a \geq 0$, $b \geq 0$, $0 \leq u \leq v$ and $n \geq 1$,*

$$\mathbf{P}\left\{ \underline{S}_n \geq -a, \; \min_{\lambda n < i \leq n} S_i \geq b - a, \; S_n \in [b - a + u, b - a + v] \right\}$$

$$\leq c_{68} \, \frac{(v + 1)(v - u + 1)(a + 1)}{n^{3/2}}. \tag{A.10}$$

Proof We treat λn as an integer. Let $\mathbf{P}_{(A.10)}$ denote the probability expression on the left-hand side of (A.10). Applying the Markov property at time λn, we see that $\mathbf{P}_{(A.10)} = \mathbf{E}[\mathbf{1}_{\{\underline{S}_{\lambda n} \geq -a, \; S_{\lambda n} \geq b-a\}} f(S_{\lambda n})]$, where $f(r) := \mathbf{P}\{\underline{S}_{n - \lambda n} \geq b - a - r, \; S_{n - \lambda n} \in [b - a - r + u, b - a - r + v]\}$ (for $r \geq b - a$). By Lemma A.2, $f(r) \leq c_{63} \frac{(v+1)(v-u+1)(a+r-b+1)}{n^{3/2}}$ (for $r \geq b - a$). Therefore,

$$\mathbf{P}_{(A.10)} \leq \frac{c_{63}(v + 1)(v - u + 1)}{n^{3/2}} \, \mathbf{E}[(S_{\lambda n} + a - b + 1) \, \mathbf{1}_{\{\underline{S}_{\lambda n} \geq -a, \; S_{\lambda n} \geq b-a\}}].$$

The expectation $\mathbf{E}[\cdots]$ on the right-hand side is bounded by $\mathbf{E}[|S_{\lambda n}| \, \mathbf{1}_{\{\underline{S}_{\lambda n} \geq -a\}}] + a + 1$, so it suffices to apply Lemma A.3.   □

Lemma A.5 *There exists a constant $c_{69} > 0$ such that for any $y \geq 0$ and $z \geq 0$,*

$$\sum_{k=0}^{\infty} \mathbf{P}\left\{ S_k \leq y - z, \ \underline{S}_k \geq -z \right\} \leq c_{69} \, (y + 1)(\min\{y, z\} + 1).$$

Proof The proof requires to apply Lemma A.1, with some care. We distinguish two possible situations.

First situation: $y < z$. Let $\tau_y^- := \inf\{i \geq 0 : S_i \leq y\}$. Then

$$\sum_{k=0}^{\infty} \mathbf{P}\{S_k \leq y - z, \ \underline{S}_k \geq -z\} = \mathbf{E}_z\left[\sum_{k=0}^{\infty} \mathbf{1}_{\{S_k \leq y, \ \underline{S}_k \geq 0\}} \right]$$

$$= \mathbf{E}_z\left[\sum_{k=\tau_y^-}^{\infty} \mathbf{1}_{\{S_k \leq y, \ \underline{S}_k \geq 0\}} \right].$$

Applying the strong Markov property at time τ_y^- gives

$$\sum_{k=0}^{\infty} \mathbf{P}\{S_k \leq y - z, \ \underline{S}_k \geq -z\} \leq \mathbf{E}\left[\sum_{i=0}^{\infty} \mathbf{1}_{\{S_i \leq y, \ \underline{S}_i \geq -y\}} \right]$$

$$= \sum_{i=0}^{\infty} \mathbf{P}(S_i \leq y, \ \underline{S}_i \geq -y).$$

For $i \leq \lfloor y^2 \rfloor$, we simply argue that the probability on the right-hand side is bounded by one. For $i \geq \lfloor y^2 \rfloor + 1$, we apply Lemma A.1 to see that the probability is bounded by $c_{70} \frac{(y+1)^3}{i^{3/2}}$. Consequently,

$$\sum_{k=0}^{\infty} \mathbf{P}\{S_k \leq y - z, \ \underline{S}_k \geq -z\} \leq (\lfloor y^2 \rfloor + 1) + \sum_{i=\lfloor y^2 \rfloor + 1}^{\infty} c_{70} \frac{(y+1)^3}{i^{3/2}},$$

which is bounded by $c_{71} (y + 1)^2$, as desired.

Second and last situation: $y \geq z$. For $k \leq \lfloor y^2 \rfloor$, we simply say $\mathbf{P}\{S_k \leq y - z, \ \underline{S}_k \geq -z\} \leq \mathbf{P}\{\underline{S}_k \geq -z\}$, which is bounded by $c_{72} \frac{z+1}{(k+1)^{1/2}}$. For $k \geq \lfloor y^2 \rfloor + 1$, we apply Lemma A.1 and obtain: $\mathbf{P}\{S_k \leq y - z, \ \underline{S}_k \geq -z\} \leq c_{62} \frac{(y+1)^2(z+1)}{k^{3/2}}$. As such,

$$\sum_{k=0}^{\infty} \mathbf{P}\{S_k \leq y - z, \ \underline{S}_k \geq -z\} \leq \sum_{k=0}^{\lfloor y^2 \rfloor} \frac{c_{72}(1 + z)}{(k + 1)^{1/2}} + \sum_{k=\lfloor y^2 \rfloor + 1}^{\infty} \frac{c_{63}(y + 1)^2(z + 1)}{k^{3/2}},$$

which is bounded by $c_{73}(y + 1)(z + 1)$. $\qquad\qquad\qquad\qquad\qquad\qquad\qquad\square$

Lemma A.6 *Let $\varepsilon > 0$ and $c_{74} > 0$. There exist constants $c_{75} > 0$ and $c_{76} > 0$ such that for all $u \geq 0$, $v \geq 0$, $a \geq 0$ and integers $n \geq 1$,*

$$\mathbf{P}_v\Big\{\exists 0 \leq i \leq n : S_i \leq k_i - c_{75}, \min_{0 \leq j \leq n} S_j \geq 0, \min_{\frac{n}{2} < j \leq n} S_j \geq a, S_n \leq a + u\Big\}$$

$$\leq (1+u)^2(1+v)\left(\frac{\varepsilon}{n^{3/2}} + c_{76}\frac{(n^{1/7}+a)^2}{n^{2-(1/7)}}\right),$$

where $k_i := c_{74}\, i^{1/7}$ for $0 \leq i \leq \lfloor\frac{n}{2}\rfloor$, and $k_i := a + c_{74}\,(n-i)^{1/7}$ for $\lfloor\frac{n}{2}\rfloor < i \leq n$.

Proof Again, we treat $\frac{n}{2}$ as an integer. We start with the trivial observation that $\{S_i \leq i^{1/7} - c_{75}\} \subset \{S_i \leq i^{1/7}\}$, regardless of the forthcoming constant c_{75}.

Let, for $0 \leq i \leq n$,

$$E_i := \Big\{S_i \leq k_i, \min_{0 \leq j \leq n} S_j \geq 0, \min_{\frac{n}{2} < j \leq n} S_j \geq a, S_n \leq a + u\Big\}.$$

[We note that the first condition in E_i is $S_i \leq k_i$, and not $S_i \leq k_i - c_{75}$ as in the statement of the lemma.] We distinguish two possible situations.

First situation: $i \leq \frac{n}{2}$, in which case $k_i = c_{74}\,i^{1/7}$. We estimate $\mathbf{P}(E_i)$ for $1 \leq i \leq \frac{n}{2}$. Applying the Markov property at time i, and using Lemma A.4 (there is no problem in applying Lemma A.4 even when i is close to $\frac{n}{2}$, because the probability expression in Lemma A.4 is non-decreasing in λ), we get

$$\mathbf{P}(E_i) \leq c_{77}\frac{(u+1)^2}{n^{3/2}}\,\mathbf{E}_v\Big[(S_i+1)\mathbf{1}_{\{S_i \leq c_{74}\,i^{1/7},\, S_i \geq 0\}}\Big],$$

which, by Lemma A.1, is bounded by $c_{78}\frac{(u+1)^2}{n^{3/2}}\frac{(i^{1/7}+1)^3(v+1)}{i^{3/2}}$. Consequently, there exists a constant i_0 sufficiently large, which does not depend on the forthcoming constant c_{75}, such that

$$\sum_{i=i_0}^{\frac{n}{2}}\mathbf{P}(E_i) \leq \frac{\varepsilon\,(u+1)^2(v+1)}{2n^{3/2}}, \tag{A.11}$$

with $\sum_{i=i_0}^{\frac{n}{2}} := 0$ if $\frac{n}{2} < i_0$.

Second and last situation: $i > \frac{n}{2}$, in which case $k_i = a + c_{74}\,(n-i)^{1/7}$ by definition. We still apply the Markov property at time i: using Lemma A.1, it is seen that

$$\mathbf{P}(E_i) \leq \frac{c_{79}\,(u+1)^2}{(n-i+1)^{3/2}}\,\mathbf{E}_v\Big[(S_i-a+1)\mathbf{1}_{\{S_i \leq a+c_{74}\,(n-i)^{1/7},\, S_i \geq 0,\, \min_{\frac{n}{2} \leq j \leq i} S_j \geq a\}}\Big]$$

$$\leq \frac{c_{80}\,(u+1)^2}{(n-i+1)^{(3/2)-(1/7)}}\,\mathbf{P}_v\Big(S_i \leq a + c_{74}\,(n-i)^{1/7},\, S_i \geq 0,$$

$$\min_{\frac{n}{2} \leq j \leq i} S_j \geq a\Big).$$

When $i \geq \frac{2n}{3}$, this yields $\mathbf{P}(E_i) \leq c_{81} (u + 1)^2 (v + 1) \frac{(n-i+1)^{(3/7)-(3/2)}}{n^{3/2}}$ by Lemma A.4. Therefore, choosing i_1 sufficiently large (not depending on the forthcoming constant c_{75}), we have

$$\sum_{i=\frac{2n}{3}}^{n-i_1} \mathbf{P}(E_i) \leq \frac{\varepsilon (u + 1)^2 (v + 1)}{2n^{3/2}}. \tag{A.12}$$

When $\frac{n}{2} < i < \frac{2n}{3}$, we simply use

$$\mathbf{P}(E_i) \leq \frac{c_{80} (u + 1)^2}{(n - i + 1)^{(3/2)-(1/7)}} \, \mathbf{P}_v \left(a \leq S_i \leq a + c_{74} (n - i)^{1/7}, \, \underline{S}_i \geq 0 \right),$$

and the probability expression on the right-hand side is bounded by $[1 + c_{74} (n - i)^{1/7}] \, \mathbf{P}_v(a \leq S_i \leq a + c_{74} (n-i)^{1/7}, \, \underline{S}_i \geq 0)$. Using Lemma A.2 gives then $\mathbf{P}(E_i) \leq c_{82} \frac{(u+1)^2(v+1)\,n^{1/7}\,(n^{1/7}+a)^2}{n^3}$; as such,

$$\sum_{i=\frac{n}{2}+1}^{\frac{2n}{3}-1} \mathbf{P}(E_i) \leq c_{83} \frac{(u + 1)^2 (v + 1) (n^{1/7} + a)^2}{n^{2-(1/7)}}.$$

Combining this with inequalities in (A.11) and (A.12), we obtain:

$$\sum_{i=i_0}^{n-i_1} \mathbf{P}(E_i) \leq \frac{\varepsilon (u + 1)^2 (v + 1)}{n^{3/2}} + c_{83} \frac{(u + 1)^2 (v + 1) (n^{1/7} + a)^2}{n^{2-(1/7)}}.$$

We choose $c_{75} > c_{74} (\max\{i_0, i_1\})^{1/7}$ so that for $i < i_0$ or for $n - i_1 < i \leq n$, the event $\{a \leq S_i \leq k_i - c_{75}\}$ is empty. The lemma is proved.

We note that the choice of $\frac{1}{7}$ in the lemma is arbitrary; any value in $(0, \frac{1}{6})$ is fine. $\qquad\square$

Lemma A.7 *There exists a constant $C_0 > 0$ such that for any $0 < a_1 \leq a_2 < \infty$,*

$$\liminf_{n\to\infty} n^{1/2} \inf_{u\in[a_1\, n^{1/2},\, a_2\, n^{1/2}]} \mathbf{P}\left\{ u \leq S_n < u + C_0 \, \Big| \, \underline{S}_n \geq 0 \right\} > 0.$$

Proof The lemma follows immediately from a conditional local limit theorem [82]: if the distribution of S_1 is non-lattice (i.e., not supported in any $a\mathbb{Z} + b$, with $a > 0$ and $b \in \mathbb{R}$), then for any $h > 0$, $\mathbf{P}\{r \leq S_n \leq r+h \,|\, \underline{S}_n \geq 0\} = \frac{hr_+}{n\mathbf{E}(S_1^2)} \exp(-\frac{r^2}{2n\mathbf{E}(S_1^2)}) + o(\frac{1}{n^{1/2}})$, $n \to \infty$, uniformly in $r \in \mathbb{R}$ (notation recalled: $r_+ := \max\{r, 0\}$); if the distribution of S_1 is lattice, and is supported in $a + b\mathbb{Z}$ with $b > 0$ being the largest such value (called the "span" in the literature), then $\mathbf{P}\{S_n = an + b\ell \,|\, \underline{S}_n \geq 0\} = \frac{b(an+b\ell)_+}{n\mathbf{E}(S_1^2)} \exp(-\frac{(an+b\ell)^2}{2n\mathbf{E}(S_1^2)}) + o(\frac{1}{n^{1/2}})$, $n \to \infty$, uniformly in $\ell \in \mathbb{Z}$. $\qquad\square$

Lemma A.8 *Let $C_0 > 0$ be the constant in Lemma A.7. For all $0 < a_1 \leq a_2 < \infty$,*

$$\liminf_{n\to\infty} n \inf_{u\in[a_1 n^{1/2}, a_2 n^{1/2}]} \mathbf{P}\left\{\underline{S}_n \geq 0, \ u \leq S_n < u + C_0\right\} > 0.$$

Proof This is a consequence of Lemma A.7 and (A.7). □

Lemma A.9 *Let $C_0 > 0$ be the constant in Lemma A.7. For all $0 < a_1 \leq a_2 < \infty$ and $a_3 > 0$,*

$$\liminf_{n\to\infty} \inf_{v\in[0, a_3 n^{1/2}]} \frac{n}{v+1} \inf_{u\in[a_1 n^{1/2}, a_2 n^{1/2}]} \mathbf{P}\left\{\underline{S}_n \geq -v, \ u \leq S_n < u + C_0\right\} > 0.$$

Proof Since $\{\underline{S}_n \geq -v\} \supset \{\underline{S}_n \geq 0\}$ for $v \geq 0$, and in view of Lemma A.8, we only need to treat the case $v \in [v_0, a_3 n^{1/2}]$ for any given $v_0 > 0$. Let $(H_k^-, k \geq 0)$ (resp. $(\theta_k^-, k \geq 0)$) be the strictly descending ladder heights (resp. ladder epochs) of (S_n), as in (A.3). We have

$$\mathbf{P}\{\underline{S}_n \geq -v, \ u \leq S_n < u + C_0\}$$

$$\geq \sum_{k=1}^{\infty} \mathbf{P}\{\theta_k^- \leq \frac{n}{2}, \ H_k^- \geq -v, \ \theta_{k+1}^- > n, \ \underline{S}_n \geq -v, \ u \leq S_n < u + C_0\}.$$

For any k, applying the strong Markov property at time θ_k^- and Lemma A.8, we see that for some $c_{84} > 0$ and $n_0 > 0$ (depending on (a_1, a_2, a_3) but not on v_0) and all $n \geq n_0$, the probability expression on the right-hand side is

$$\geq \mathbf{P}\{\theta_k^- \leq \frac{n}{2}, \ H_k^- \geq -v\} \times \frac{c_{84}}{n};$$

hence

$$\mathbf{P}\{\underline{S}_n \geq -v, \ u \leq S_n < u + C_0\} \geq \frac{c_{84}}{n} \sum_{k=1}^{\infty} \mathbf{P}\{\theta_k^- \leq \frac{n}{2}, \ H_k^- \geq -v\}$$

$$\geq \frac{c_{84}}{n} k_0 \, \mathbf{P}\{\theta_{k_0}^- \leq \frac{n}{2}, \ H_{k_0}^- \geq -v\},$$

for any $k_0 \geq 1$. Since $\frac{|H_k^-|}{k} \to \frac{1}{c_{\text{ren}}} \in (0, \infty)$ a.s. (for $k \to \infty$; see (A.4) and (A.5)), and θ_k^- is the first time the random walk (S_i) hits $(-\infty, H_k^-]$, it follows from Donsker's theorem that we can choose v_0 sufficiently large and $\delta > 0$ sufficiently small, such that with the choice of $k_0 := \lfloor \delta v \rfloor$, we have, for all $v \in [v_0, a_3 n^{1/2}]$ and all sufficiently large n, $\mathbf{P}\{H_{k_0}^- \geq -v\} \geq \frac{2}{3}$ and $\mathbf{P}\{\theta_{k_0}^- \leq \frac{n}{2}\} \geq \frac{2}{3}$, so that $\mathbf{P}\{\theta_{k_0}^- \leq \frac{n}{2}, \ H_{k_0}^- \geq -v\} \geq \frac{2}{3} + \frac{2}{3} - 1 = \frac{1}{3}$. The lemma follows. □

Lemma A.10 *There exists a constant $C > 0$ such that for all $a_1 > 0$, $a_2 > 0$ and $0 < \lambda < 1$, there exist an integer $n_0 \geq 1$ and a constant $c_{85} > 0$ such that the* *following inequality holds:*

$$\mathbf{P}\left\{\underline{S}_{\lfloor\lambda n\rfloor} \geq -v, \min_{\lfloor\lambda n\rfloor < j \leq n} S_j \geq u, \ S_n \leq u + C\right\} \geq c_{85}\,\frac{v+1}{n^{3/2}}, \qquad (A.13)$$

for all $n \geq n_0$, all $u \in [0, a_1 n^{1/2}]$ and all $v \in [0, a_2 n^{1/2}]$.

Proof We treat λn and $n^{1/2}$ as integers. Let $C_0 > 0$ be the constant in Lemma A.7, and we take $C := 2C_0$. Let $\mathbf{P}_{(A.13)}$ denote the probability on the left-hand side of (A.13). Writing $\alpha_k := 2a_1 n^{1/2} + kC_0$ for $k \geq 0$, we have

$$\mathbf{P}_{(A.13)} \geq \sum_{k=0}^{n^{1/2}} \mathbf{P}\Big\{\underline{S}_{\lambda n} \geq -v, \ \alpha_k \leq S_{\lambda n} < \alpha_{k+1}, \ \min_{\lambda n < j \leq n}(S_j - S_{\lambda n}) \geq u - \alpha_k,$$

$$S_n - S_{\lambda n} \leq u + 2C_0 - \alpha_{k+1}\Big\}$$

$$= \sum_{k=0}^{n^{1/2}} \mathbf{P}\Big\{\underline{S}_{\lambda n} \geq -v, \ \alpha_k \leq S_{\lambda n} < \alpha_{k+1}\Big\}$$

$$\times \mathbf{P}\Big\{\underline{S}_{(1-\lambda)n} \geq u - \alpha_k, \ S_{(1-\lambda)n} \leq u + C_0 - \alpha_k\Big\}.$$

[In the last identity, we have used the fact that $u + 2C_0 - \alpha_{k+1} = u + C_0 - \alpha_k$.] We need to treat the two probability expressions on the right-hand side. The first is easy: by Lemma A.9, there exists a constant $c_{86} > 0$ such that for all sufficiently large n,

$$\min_{0 \leq k \leq n^{1/2}} \mathbf{P}\Big\{\underline{S}_{\lambda n} \geq -v, \ \alpha_k \leq S_{\lambda n} < \alpha_{k+1}\Big\} \geq c_{86}\,\frac{v+1}{n}.$$

To treat the second probability expression on the right-hand side, we write $\widehat{S}_j := S_{(1-\lambda)n-j} - S_{(1-\lambda)n}$, to see that

$$\mathbf{P}\Big\{\underline{S}_{(1-\lambda)n} \geq u - \alpha_k, \ S_{(1-\lambda)n} \leq u + C_0 - \alpha_k\Big\}$$

$$\geq \mathbf{P}\Big\{\min_{1 \leq j \leq (1-\lambda)n} \widehat{S}_j \geq 0, \ -u - C_0 + \alpha_k \leq \widehat{S}_{(1-\lambda)n} \leq -u + \alpha_k\Big\},$$

which, by Lemma A.8, is greater than $\frac{c_{87}}{n}$ for some $c_{87} > 0$ and all sufficiently large n, uniformly in $0 \leq k \leq n^{1/2}$. Consequently, for all sufficiently large n,

$$\mathbf{P}_{(A.13)} \geq \sum_{k=0}^{n^{1/2}} c_{86}\,\frac{v+1}{n}\,\frac{c_{87}}{n},$$

which is greater than $c_{88}\,\frac{v+1}{n^{3/2}}$ for some constant $c_{88} > 0$. $\square$

References

1. L. Addario-Berry, N. Broutin, Total progeny in killed branching random walk. Probab. Theory Relat. Fields **151**, 265–295 (2011)
2. L. Addario-Berry, B.A. Reed, Ballot theorems, old and new, in *Horizons of Combinatorics*. Bolyai Society Mathematical Studies, vol. 17 (Springer, Berlin, 2008), pp. 9–35
3. L. Addario-Berry, B.A. Reed, Minima in branching random walks. Ann. Probab. **37**, 1044–1079 (2009)
4. L. Addario-Berry, L. Devroye, S. Janson, Sub-Gaussian tail bounds for the width and height of conditioned Galton–Watson trees. Ann. Probab. **41**, 1072–1087 (2013)
5. E. Aïdékon, Transient random walks in random environment on a Galton–Watson tree. Probab. Theory Relat. Fields **142**, 525–559 (2008)
6. E. Aïdékon, Large deviations for transient random walks in random environment on a Galton–Watson tree. Ann. Inst. H. Poincaré Probab. Stat. **46**, 159–189 (2010)
7. E. Aïdékon, Tail asymptotics for the total progeny of the critical killed branching random walk. Electron. Commun. Probab. **15**, 522–533 (2010)
8. E. Aïdékon, Convergence in law of the minimum of a branching random walk. Ann. Probab. **41**, 1362–1426 (2013). ArXiv version `ArXiv:1101.1810`
9. E. Aïdékon, Monotonicity for $\lambda \leq \frac{1}{2}$ (2013+). Available at: http://www.proba.jussieu.fr/~aidekon (Preprint)
10. E. Aïdékon, Speed of the biased random walk on a Galton–Watson tree. Probab. Theory Relat. Fields **159**, 597–617 (2014)
11. E. Aïdékon, The extremal process in nested conformal loops (2015+). Preprint available at: http://www.proba.jussieu.fr/~aidekon
12. E. Aïdékon, S.C. Harris, Near-critical survival probability of branching brownian motion with an absorbing barrier (2015+, in preparation)
13. E. Aïdékon, B. Jaffuel, Survival of branching random walks with absorption. Stoch. Process. Appl. **121**, 1901–1937 (2011)
14. E. Aïdékon, Z. Shi, Weak convergence for the minimal position in a branching random walk: a simple proof. Period. Math. Hung. **61**, 43–54 (2010). Special issue in honour of E. Csáki and P. Révész
15. E. Aïdékon, Z. Shi, The Seneta–Heyde scaling for the branching random walk. Ann. Probab. **42**, 959–993 (2014)
16. E. Aïdékon, J. Berestycki, Brunet, É., Z. Shi, Branching Brownian motion seen from its tip. Probab. Theory Relat. Fields **157**, 405–451 (2013)
17. E. Aïdékon, Y. Hu, O. Zindy, The precise tail behavior of the total progeny of a killed branching random walk. Ann. Probab. **41**, 3786–3878 (2013)

© Springer International Publishing Switzerland 2015
Z. Shi, *Branching Random Walks*, Lecture Notes in Mathematics 2151,
DOI 10.1007/978-3-319-25372-5

18. T. Alberts, M. Ortgiese, The near-critical scaling window for directed polymers on disordered trees. Electron. J. Probab. **18**, 1–24 (2013). Paper No. 19

19. D. Aldous, Power laws and killed branching random walks (1999). Available at: http://www.stat.berkeley.edu/~aldous/Research/OP/brw.html

20. G. Alsmeyer, The smoothing transform: a review of contraction results, in *Random Matrices and Iterated Random Functions*. Springer Proceedings in Mathematics and Statistics, vol. 53 (Springer, Heidelberg, 2013), pp. 189–228

21. G. Alsmeyer, J.D. Biggins, M. Meiners, The functional equation of the smoothing transform. Ann. Probab. **40**, 2069–2105 (2012)

22. G. Alsmeyer, E. Damek, S. Mentemeier, Precise tail index of fixed points of the two-sided smoothing transform, in *Random Matrices and Iterated Random Functions*. Springer Proceedings in Mathematics and Statistics, vol. 53 (Springer, Heidelberg, 2013), pp. 229–251

23. O. Amini, L. Devroye, S. Griffiths, N. Olver, On explosions in heavy-tailed branching random walks. Ann. Probab. **41**, 1864–1899 (2013)

24. P. Andreoletti, P. Debs, Spread of visited sites of a random walk along the generations of a branching process. Electron. J. Probab. **19**, 1–22 (2014)

25. P. Andreoletti, P. Debs, The number of generations entirely visited for recurrent random walks on random environment. J. Theor. Probab. **27**, 518–538 (2014)

26. L.-P. Arguin, Work in progress (2015+). Announced at: http://www.dms.umontreal.ca/~arguinlp/recherche.html

27. L.-P. Arguin, A. Bovier, N. Kistler, The genealogy of extremal particles of branching Brownian motion. Commun. Pure Appl. Math. **64**, 1647–1676 (2011)

28. L.-P. Arguin, A. Bovier, N. Kistler, Poissonian statistics in the extremal process of branching Brownian motion. Ann. Appl. Probab. **22**, 1693–1711 (2012)

29. L.-P. Arguin, A. Bovier, N. Kistler, An ergodic theorem for the frontier of branching Brownian motion. Electron. J. Probab. **18**, 1–25 (2013). Paper No. 53

30. L.-P. Arguin, A. Bovier, N. Kistler, The extremal process of branching Brownian motion. Probab. Theory Relat. Fields **157**, 535–574 (2013)

31. S. Asmussen, H. Hering, *Branching Processes* (Birkhäuser, Basel, 1983)

32. K.B. Athreya, P.E. Ney, *Branching Processes* (Springer, New York, 1972)

33. M. Bachmann, Limit theorems for the minimal position in a branching random walk with independent logconcave displacements. Adv. Appl. Probab. **32**, 159–176 (2000)

34. J. Barral, Y. Hu, T. Madaule, The minimum of a branching random walk outside the boundary case (2014+). ArXiv:1406.6971

35. R.F. Bass, P.S. Griffin, The most visited site of Brownian motion and simple random walk. Z. Wahrscheinlichkeitstheorie verw. Gebiete **70**, 417–436 (1985)

36. D. Belius, N. Kistler, The subleading order of two dimensional cover times (2014+). ArXiv:1405.0888

37. G. Ben Arous, A. Fribergh, N. Gantert, A. Hammond, Biased random walks on a Galton-Watson tree with leaves. Ann. Probab. **40**, 280–338 (2012)

38. G. Ben Arous, Y. Hu, S. Olla, O. Zeitouni, Einstein relation for biased random walk on Galton–Watson trees. Ann. Inst. H. Poincaré Probab. Stat. **49**, 698–721 (2013)

39. G. Ben Arous, A. Fribergh, V. Sidoravicius, Lyons-Pemantle-Peres monotonicity problem for high biases. Commun. Pure Appl. Math. **67**, 519–530 (2014)

40. J. Bérard, J.-B. Gouéré, Brunet-Derrida behavior of branching-selection particle systems on the line. Commun. Math. Phys. **298**, 323–342 (2010)

41. J. Bérard, J.-B. Gouéré, Survival probability of the branching random walk killed below a linear boundary. Electron. J. Probab. **16**, 396–418 (2011). Paper No. 14

42. J. Bérard, P. Maillard, The limiting process of N-particle branching random walk with polynomial tails. Electron. J. Probab. **19**, 1–17 (2014). Paper No. 22

43. J. Berestycki, Topics on branching Brownian motion (2015+). Lecture notes available at: http://www.stats.ox.ac.uk/~berestyc/articles.html

44. N. Berestycki, L.Z. Zhao, The shape of multidimensional Brunet–Derrida particle systems (2013+). ArXiv:1305.0254

45. J. Berestycki, N. Berestycki, J. Schweinsberg, Survival of near-critical branching Brownian motion. J. Stat. Phys. **143**, 833–854 (2011)
46. J. Berestycki, N. Berestycki, J. Schweinsberg, The genealogy of branching Brownian motion with absorption. Ann. Probab. **41**, 527–618 (2013)
47. J. Bertoin, *Random Fragmentation and Coagulation Processes* (Cambridge University Press, Cambridge, 2006)
48. P. Biane, M. Yor, Valeurs principales associées aux temps locaux browniens. Bull. Sci. Math. **111**, 23–101 (1987)
49. J.D. Biggins, The first- and last-birth problems for a multitype age-dependent branching process. Adv. Appl. Probab. **8**, 446–459 (1976)
50. J.D. Biggins, Martingale convergence in the branching random walk. J. Appl. Probab. **14**, 25–37 (1977)
51. J.D. Biggins, Uniform convergence of martingales in the branching random walk. Ann. Probab. **20**, 137–151 (1992)
52. J.D. Biggins, Lindley-type equations in the branching random walk. Stoch. Process. Appl. **75**, 105–133 (1998)
53. J.D. Biggins, Random walk conditioned to stay positive. J. Lond. Math. Soc. **67**, 259–272 (2003)
54. J.D. Biggins, Branching out, in *Probability and Mathematical Genetics: Papers in Honour of Sir John Kingman*, ed. by N.H. Bingham, C.M. Goldie (Cambridge University Press, Cambridge, 2010), pp. 112–133
55. J.D. Biggins, D.R. Grey, Continuity of limit random variables in the branching random walk. J. Appl. Probab. **16**, 740–749 (1979)
56. J.D. Biggins, A.E. Kyprianou, Seneta-Heyde norming in the branching random walk. Ann. Probab. **25**, 337–360 (1997)
57. J.D. Biggins, A.E. Kyprianou, Measure change in multitype branching. Adv. Appl. Probab. **36**, 544–581 (2004)
58. J.D. Biggins, A.E. Kyprianou, Fixed points of the smoothing transform: the boundary case. Electron. J. Probab. **10**, 609–631 (2005). Paper No. 17
59. J.D. Biggins, B.D. Lubachevsky, A. Shwartz, A. Weiss, A branching random walk with barrier. Ann. Appl. Probab. **1**, 573–581 (1991)
60. M. Biskup, O. Louidor, Conformal symmetries in the extremal process of two-dimensional discrete Gaussian Free Field (2014+). `ArXiv:1410.4676`
61. E. Bolthausen, J.-D. Deuschel, G. Giacomin, Entropic repulsion and the maximum of the two-dimensional harmonic crystal. Ann. Probab. **29**, 1670–1692 (2001)
62. A.A. Borovkov, S.G. Foss, Estimates for overshooting an arbitrary boundary by a random walk and their applications. Theory Probab. Appl. **44**, 231–253 (2000)
63. A. Bovier, L. Hartung, The extremal process of two-speed branching Brownian motion. Electron. J. Probab. **19**, 1–28 (2014). Paper No. 18
64. A. Bovier, L. Hartung, Extended convergence of the extremal process of branching Brownian motion (2014+). `ArXiv:1412.5975`
65. A. Bovier, L. Hartung, Variable speed branching Brownian motion 1. Extremal processes in the weak correlation regime. ALEA Lat. Am. J. Probab. Math. Stat. **12**, 261–291 (2015)
66. A. Bovier, I. Kurkova, Derrida's generalized random energy models 2: models with continuous hierarchies. Ann. Inst. H. Poincaré Probab. Stat. **40**, 481–495 (2004)
67. M.D. Bramson, Maximal displacement of branching Brownian motion. Commun. Pure Appl. Math. **31**, 531–581 (1978)
68. M.D. Bramson, Maximal displacement of branching random walk. Z. Wahrsch. Gebiete **45**, 89–108 (1978)
69. M.D. Bramson, Convergence of solutions of the Kolmogorov equation to travelling waves. Mem. Am. Math. Soc. **44**(285), iv + 190pp. (1983)
70. M.D. Bramson, O. Zeitouni, Tightness for a family of recursion equations. Ann. Probab. **37**, 615–653 (2009)

71. M. Bramson, J. Ding, O. Zeitouni, Convergence in law of the maximum of nonlattice branching random walk (2014+). ArXiv:1404.3423

72. H.P. Breuer, W. Huber, F. Petruccione, Fluctuation effects on wave propagation in a reaction-diffusion process. Physica D **73**, 259–273 (1994)

73. É. Brunet, B. Derrida, Shift in the velocity of a front due to a cutoff. Phys. Rev. E **56**, 2597–2604 (1997)

74. É. Brunet, B. Derrida, Effect of microscopic noise on front propagation. J. Stat. Phys. **103**, 269–282 (2001)

75. É. Brunet, B. Derrida, Statistics at the tip of a branching random walk and the delay of traveling waves. Europhys. Lett. **87**, 60010 (2009)

76. É. Brunet, B. Derrida, A branching random walk seen from the tip. J. Stat. Phys. **143**, 420–446 (2011)

77. É. Brunet, B. Derrida, A.H. Mueller, S. Munier, A phenomenological theory giving the full statistics of the position of fluctuating pulled fronts. Phys. Rev. E **73**, 056126 (2006)

78. É. Brunet, B. Derrida, A.H. Mueller, S. Munier, Noisy traveling waves: effect of selection on genealogies. Europhys. Lett. **76**, 1–7 (2006)

79. É. Brunet, B. Derrida, A.H. Mueller, S. Munier, Effect of selection on ancestry: an exactly soluble case and its phenomenological generalization. Phys. Rev. E **76**, 041104 (2007)

80. D. Buraczewski, On tails of fixed points of the smoothing transform in the boundary case. Stoch. Process. Appl. **119**, 3955–3961 (2009)

81. D. Buraczewski, E. Damek, S. Mentemeier, M. Mirek, Heavy tailed solutions of multivariate smoothing transforms. Stoch. Process. Appl. **123**, 1947–1986 (2013)

82. F. Caravenna, A local limit theorem for random walks conditioned to stay positive. Probab. Theory Relat. Fields **133**, 508–530 (2005)

83. F. Caravenna, L. Chaumont, Invariance principles for random walks conditioned to stay positive. Ann. Inst. H. Poincaré Probab. Stat. **44**, 170–190 (2008)

84. B. Chauvin, Product martingales and stopping lines for branching Brownian motion. Ann. Probab. **19**, 1195–1205 (1991)

85. B. Chauvin, A. Rouault, KPP equation and supercritical branching Brownian motion in the subcritical speed area. Application to spatial trees. Probab. Theory Relat. Fields **80**, 299–314 (1988)

86. D. Chen, Average properties of random walks on Galton–Watson trees. Ann. Inst. H. Poincaré Probab. Stat. **33**, 359–369 (1997)

87. X. Chen, Scaling limit of the path leading to the leftmost particle in a branching random walk (2013+). ArXiv:1305.6723

88. X. Chen, A necessary and sufficient condition for the non-trivial limit of the derivative martingale in a branching random walk (2014+). ArXiv:1402.5864

89. F. Comets, Directed polymers in random environment, in *Lecture Notes. Random Interfaces and Directed Polymers*, Leipzig, 11–17 Sept 2005

90. F.M. Dekking, B. Host, Limit distributions for minimal displacement of branching random walks. Probab. Theory Relat. Fields **90**, 403–426 (1991)

91. A. Dembo, N. Gantert, Y. Peres, O. Zeitouni, Large deviations for random walks on Galton–Watson trees: averaging and uncertainty. Probab. Theory Relat. Fields **122**, 241–288 (2002)

92. B. Derrida, Z. Shi, Large deviations for the branching Brownian motion in presence of selection or coalescence (2015+, in preparation)

93. B. Derrida, D. Simon, The survival probability of a branching random walk in presence of an absorbing wall. Europhys. Lett. **78**, 60006 (2007)

94. B. Derrida, H. Spohn, Polymers on disordered trees, spin glasses, and traveling waves. J. Stat. Phys. **51**, 817–840 (1988)

95. J. Ding, On cover times for 2D lattices. Electron. J. Probab. **17**, 1–18 (2012). Paper No. 45

96. J. Ding, R. Roy, O. Zeitouni, Convergence of the centered maximum of log-correlated Gaussian fields (2015+). ArXiv:1503.04588

97. C.R. Doering, C. Mueller, P. Smereka, Interacting particles, the stochastic Fisher–Kolmogorov–Petrovsky–Piscounov equation, and duality. Physica A **325**, 243–259 (2003)

98. R.A. Doney, A limit theorem for a class of supercritical branching processes. J. Appl. Probab. **9**, 707–724 (1972)
99. R.A. Doney, Moments of ladders heights in random walks. J. Appl. Probab. **17**, 248–252 (1980)
100. M. Drmota, *Random Trees. An Interplay Between Combinatorics and Probability* (Springer, Vienna, 2009)
101. B. Duplantier, R. Rhodes, S. Sheffield, V. Vargas, Log-correlated Gaussian fields: an overview (2014+). ArXiv:1407.5605
102. R. Durrett, *Probability: Theory and Examples*, 4th edn. (Cambridge University Press, Cambridge, 2010)
103. R. Durrett, T.M. Liggett, Fixed points of the smoothing transformation. Z. Wahrsch. Gebiete **64**, 275–301 (1983)
104. R. Durrett, D. Remenik, Brunet-Derrida particle systems, free boundary problems and Wiener-Hopf equations. Ann. Probab. **39**, 2043–2078 (2011)
105. P. Erdős, P. Révész, On the favourite points of a random walk, in *Mathematical Structures – Computational Mathematics – Mathematical Modelling*, vol. 2 (Sofia, Manchester, NH, 1984), pp. 152–157
106. K.J. Falconer, Cut-set sums and tree processes. Proc. Am. Math. Soc. **101**, 337–346 (1987)
107. M. Fang, O. Zeitouni, Consistent minimal displacement of branching random walks. Electron. Commun. Probab. **15**, 106–118 (2010)
108. M. Fang, O. Zeitouni, Branching random walks in time inhomogeneous environments. Electron. J. Probab. **17** (2012). Paper No. 67
109. M. Fang, O. Zeitouni, Slowdown for time inhomogeneous branching Brownian motion. J. Stat. Phys. **149**, 1–9 (2012)
110. G. Faraud, A central limit theorem for random walk in random environment on marked Galton-Watson trees. Electron. J. Probab. **16**, 174–215 (2011)
111. G. Faraud, Y. Hu, Z. Shi, Almost sure convergence for stochastically biased random walks on trees. Probab. Theory Relat. Fields **154**, 621–660 (2012)
112. W. Feller, *An Introduction to Probability Theory and Its Applications II*, 2nd edn. (Wiley, New York, 1971)
113. R.A. Fisher, The wave of advance of advantageous genes. Ann. Human Genet. **7**, 355–369 (1937)
114. N. Gantert, The maximum of a branching random walk with semiexponential increments. Ann. Probab. **28**, 1219–1229 (2000)
115. N. Gantert, Y. Hu, Z. Shi, Asymptotics for the survival probability in a killed branching random walk. Ann. Inst. H. Poincaré Probab. Stat. **47**, 111–129 (2011)
116. C. Garban, R. Rhodes, V. Vargas, Liouville Brownian motion (2013+). ArXiv:1301.2876
117. J.-B. Gouéré, Le mouvement Brownien branchant vu depuis sa particule la plus à gauche (d'après Arguin-Bovier-Kistler et Aïdékon-Berestycki-Brunet-Shi). Séminaire Bourbaki Astérisque **361**, 271–298 (2014). Exposé numéro 1067
118. T. Gross, Marche aléatoire en milieu aléatoire sur un arbre. Thèse de doctorat de l'Université Paris Diderot, 2004
119. Y. Guivarc'h, Sur une extension de la notion de loi semi-stable. Ann. Inst. H. Poincaré Probab. Stat. **26**, 261–285 (1990)
120. J.M. Hammersley, Postulates for subadditive processes. Ann. Probab. **2**, 652–680 (1974)
121. A. Hammond, Stable limit laws for randomly biased walks on supercritical trees. Ann. Probab. **41**, 1694–1766 (2013)
122. T.E. Harris, *The Theory of Branching Processes* (Springer, Berlin, 1963)
123. S.C. Harris, Travelling-waves for the FKPP equation via probabilistic arguments. Proc. R. Soc. Edinb. Sect. A **129**, 503–517 (1999)
124. J.W. Harris, S.C. Harris, Survival probabilities for branching Brownian motion with absorption. Electron. Commun. Probab. **12**, 89–100 (2007)
125. S.C. Harris, M.I. Roberts, Measure changes with extinction. Stat. Probab. Lett. **79**, 1129–1133 (2009)

126. S.C. Harris, M.I. Roberts, The many-to-few lemma and multiple spines (2011+). ArXiv:1106.4761

127. J.W. Harris, S.C. Harris, A.E. Kyprianou, Further probabilistic analysis of the Fisher-Kolmogorov-Petrovskii-Piscounov equation: one sided travelling waves. Ann. Inst. H. Poincaré Probab. Stat. **42**, 125–145 (2006)

128. S.C. Harris, M. Hesse, A.E. Kyprianou, Branching Brownian motion in strip: survival near criticality (2012+). ArXiv:1212.1444

129. C.C. Heyde, Extension of a result of Seneta for the super-critical Galton–Watson process. Ann. Math. Stat. **41**, 739–742 (1970)

130. R. Holley, T.M. Liggett, Generalized potlatch and smoothing processes. Z. Wahrsch. Gebiete **55**, 165–195 (1981)

131. Y. Hu, The almost sure limits of the minimal position and the additive martingale in a branching random walk (2012+). ArXiv:1211.5309

132. Y. Hu, How big is the minimum of a branching random walk? (2013+). ArXiv:1305.6448

133. Y. Hu, Local times of subdiffusive biased walks on trees (2014+). ArXiv:1412.4507

134. Y. Hu, Z. Shi, The problem of the most visited site in random environment. Probab. Theory Relat. Fields **116**, 273–302 (2000)

135. Y. Hu, Z. Shi, A subdiffusive behaviour of recurrent random walk in random environment on a regular tree. Probab. Theory Relat. Fields **138**, 521–549 (2007)

136. Y. Hu, Z. Shi, Slow movement of recurrent random walk in random environment on a regular tree. Ann. Probab. **35**, 1978–1997 (2007)

137. Y. Hu, Z. Shi, Minimal position and critical martingale convergence in branching random walks, and directed polymers on disordered trees. Ann. Probab. **37**, 742–789 (2009)

138. Y. Hu, Z. Shi, The most visited sites of biased random walks on trees. Electron. J. Probab. **20**, 1–14 (2015). Paper No. 62

139. Y. Hu, Z. Shi, The slow regime of randomly biased walks on trees (2015+). ArXiv:1501.07700

140. Y. Hu, Z. Shi, M. Yor, The maximal drawdown of the Brownian meander. Electron. Commun. Probab. **20**, 1–6 (2015). Paper No. 39

141. N. Ikeda, M. Nagasawa, S. Watanabe, Branching Markov processes I. J. Math. Kyoto Univ. **8**, 233–278 (1968)

142. N. Ikeda, M. Nagasawa, S. Watanabe, Branching Markov processes II. J. Math. Kyoto Univ. **8**, 365–410 (1968)

143. N. Ikeda, M. Nagasawa, S. Watanabe, Branching Markov processes III. J. Math. Kyoto Univ. **9**, 95–160 (1969)

144. A. Iksanov, Z. Kabluchko, A central limit theorem and a law of the iterated logarithm for the Biggins martingale of the supercritical branching random walk (2015+). ArXiv:1507.08458

145. A. Iksanov, M. Meiners, Exponential rate of almost-sure convergence of intrinsic martingales in supercritical branching random walks. J. Appl. Probab. **47**, 513–525 (2010)

146. B. Jaffuel, The critical random barrier for the survival of branching random walk with absorption. Ann. Inst. H. Poincaré Probab. Stat. **48**, 989–1009 (2012). ArXiv version ArXiv:0911.2227

147. P. Jagers, *Branching Processes with Biological Applications* (Wiley, Chichester, 1975)

148. P. Jagers, General branching processes as Markov fields. Stoch. Process. Appl. **32**, 183–212 (1989)

149. A. Joffe, A new martingale in branching random walk. Ann. Appl. Probab. **3**, 1145–1150 (1993)

150. J.-P. Kahane, Sur le modèle de turbulence de Benoît Mandelbrot. C. R. Acad. Sci. Paris Sér. A **278**, 621–623 (1974)

151. J.-P. Kahane, Sur le chaos multiplicatif. Ann. Sci. Math. Québec **9**, 105–150 (1985)

152. J.-P. Kahane, J. Peyrière, Sur certaines martingales de Mandelbrot. Adv. Math. **22**, 131–145 (1976)

153. D.A. Kessler, Z. Ner, L.M. Sander, Front propagation: precursors, cutoffs and structural stability. Phys. Rev. E **58**, 107–114 (1998)
154. H. Kesten, Branching Brownian motion with absorption. Stoch. Process. Appl. **37**, 9–47 (1978)
155. H. Kesten, B.P. Stigum, A limit theorem for multidimensional Galton–Watson processes. Ann. Math. Stat. **37**, 1211–1223 (1966)
156. J.F.C. Kingman, The first birth problem for an age-dependent branching process. Ann. Probab. **3**, 790–801 (1975)
157. A.N. Kolmogorov, S.V. Fomin, *Introductory Real Analysis* (Dover, New York, 1970)
158. A.N. Kolmogorov, I. Petrovskii, N. Piskunov, Étude de l'équation de la diffusion avec croissance de la quantité de matière et son application à un problème biologique. Bull. Univ. Moscou Série internationale, Section A, Mathématiques et mécanique **1**, 1–25 (1937)
159. M.V. Kozlov, The asymptotic behavior of the probability of non-extinction of critical branching processes in a random environment. Theory Probab. Appl. **21**, 791–804 (1976)
160. A.E. Kyprianou, Slow variation and uniqueness of solutions to the functional equation in the branching random walk. J. Appl. Probab. **35**, 795–801 (1998)
161. A.E. Kyprianou, Martingale convergence and the stopped branching random walk. Probab. Theory Relat. Fields **116**, 405–419 (2000)
162. A.E. Kyprianou, Travelling wave solutions to the K-P-P equation: alternatives to Simon Harris' probabilistic analysis. Ann. Inst. H. Poincaré Probab. Stat. **40**, 53–72 (2004)
163. T.Z. Lai, Asymptotic moments of random walks with applications to ladder variables and renewal theory. Ann. Probab. **4**, 51–66 (1976)
164. S.P. Lalley, T. Sellke, A conditional limit theorem for the frontier of a branching Brownian motion. Ann. Probab. **15**, 1052–1061 (1987)
165. M.A. Lifshits, Cyclic behavior of the maximum in a hierarchical summation scheme. J. Math. Sci. (N. Y.) **199**, 215–224 (2014)
166. T.M. Liggett, An improved subadditive ergodic theorem. Ann. Probab. **13**, 1279–1285 (1985)
167. Q.S. Liu, Fixed points of a generalized smoothing transform and applications to the branching processes. Adv. Appl. Probab. **30**, 85–112 (1998)
168. Q.S. Liu, On generalized multiplicative cascades. Stoch. Process. Appl. **86**, 263–286 (2000)
169. Q.S. Liu, Asymptotic properties and absolute continuity of laws stable by random weighted mean. Stoch. Process. Appl. **95**, 83–107 (2001)
170. R.-L. Liu, Y.-X. Ren, R. Song, $L \ln L$ criterion for a class of superdiffusions. J. Appl. Appl. **46**, 479–496 (2009)
171. R. Lyons, Random walks and percolation on trees. Ann. Probab. **18**, 931–958 (1990)
172. R. Lyons, Random walks, capacity and percolation on trees. Ann. Probab. **20**, 2043–2088 (1992)
173. R. Lyons, A simple path to Biggins' martingale convergence for branching random walk, in *Classical and Modern Branching Processes*, ed. by K.B. Athreya, P. Jagers. IMA Volumes in Mathematics and Its Applications, vol. 84 (Springer, New York, 1997), pp. 217–221
174. R. Lyons, R. Pemantle, Random walk in a random environment and first-passage percolation on trees. Ann. Probab. **20**, 125–136 (1992)
175. R. Lyons, Y. Peres, *Probability on Trees and Networks* (Cambridge University Press, Cambridge, 2015+, in preparation). Current version available at: http://mypage.iu.edu/~rdlyons/prbtree/prbtree.html
176. R. Lyons, R. Pemantle, Y. Peres, Conceptual proofs of $L \ln L$ criteria for mean behavior of branching processes. Ann. Probab. **23**, 1125–1138 (1995)
177. R. Lyons, R. Pemantle, Y. Peres, Ergodic theory on Galton–Watson trees: speed of random walk and dimension of harmonic measure. Ergodic Theory Dyn. Syst. **15**, 593–619 (1995)
178. R. Lyons, R. Pemantle, Y. Peres, Biased random walks on Galton–Watson trees. Probab. Theory Relat. Fields **106**, 249–264 (1996)
179. C. McDiarmid, Minimal positions in a branching random walk. Ann. Appl. Probab. **5**, 128–139 (1995)

180. H.P. McKean, Application of Brownian motion to the equation of Kolmogorov-Petrovskii-Piskunov. Commun. Pure Appl. Math. **28**, 323–331 (1975). Erratum: **29**, 553–554

181. T. Madaule, Convergence in law for the branching random walk seen from its tip (2011+). `ArXiv:1107.2543`

182. T. Madaule, First order transition for the branching random walk at the critical parameter (2012+). `ArXiv:1206.3835`

183. T. Madaule, Maximum of a log-correlated gaussian field (2013+). `ArXiv:1307.1365`

184. P. Maillard, A note on stable point processes occurring in branching Brownian motion. Electron. Commun. Probab. **18** (2013). Paper No. 5

185. P. Maillard, The number of absorbed individuals in branching Brownian motion with a barrier. Ann. Inst. H. Poincaré Probab. Stat. **49**, 428–455 (2013)

186. P. Maillard, Speed and fluctuations of N-particle branching Brownian motion with spatial selection (2013+). `ArXiv:1304.0562`

187. P. Maillard, O. Zeitouni, Slowdown in branching Brownian motion with inhomogeneous variance (2013+). `ArXiv:1307.3583`

188. B. Mallein, Consistent maximal displacement of the branching random walk, in *Marches aléatoires branchantes, environnement inhomogène, sélection*. Thèse de doctorat de l'Université Pierre et Marie Curie, 2015

189. B. Mallein, Maximal displacement of a branching random walk in time-inhomogeneous environment. Stoch. Process. Appl. **125**, 3958–4019 (2015)

190. B. Mallein, Maximal displacement in a branching random walk through interfaces. Electron. J. Probab. **20**, 1–40 (2015). Paper No. 68

191. B. Mallein, Branching random walk with selection at critical rate (2015+). `ArXiv:1502.07390`

192. B. Mallein, N-Branching random walk with α-stable spine (2015+). `ArXiv:1503.03762`

193. B. Mandelbrot, Multiplications aléatoires itérées et distributions invariantes par moyenne pondérée aléatoire. C. R. Acad. Sci. Paris Sér. A **278**, 289–292 (1974)

194. R.D. Mauldin, S.C. Williams, Random recursive constructions: asymptotic geometric and topological properties. Trans. Am. Math. Soc. **295**, 325–346 (1986)

195. M.V. Menshikov, D. Petritis, On random walks in random environment on trees and their relationship with multiplicative chaos, in *Mathematics and Computer Science II (Versailles, 2002)* (Birkhäuser, Basel, 2002), pp. 415–422

196. A.A. Mogulskii, Absolute estimates for moments of certain boundary functionals. Theory Probab. Appl. **18**, 340–347 (1973)

197. A.A. Mogulskii, Small deviations in the space of trajectories. Theory Probab. Appl. **19**, 726–736 (1974)

198. P. Mörters, M. Ortgiese, Minimal supporting subtrees for the free energy of polymers on disordered trees. J. Math. Phys. **49**, 125203 (2008)

199. C. Mueller, L. Mytnik, J. Quastel, Small noise asymptotics of traveling waves. Markov Proc. Relat. Fields **14**, 333–342 (2008)

200. C. Mueller, L. Mytnik, J. Quastel, Effect of noise on front propagation in reaction-diffusion equations of KPP type. Invent. Math. **184**, 405–453 (2011)

201. S. Munier, R. Peschanski, Geometric scaling as traveling waves. Phys. Rev. Lett. **91**, 232001 (2003)

202. O. Nerman, On the convergence of supercritical general (C-M-J) branching processes. Z. Wahrsch. Gebiete **57**, 365–395 (1981)

203. J. Neveu, Arbres et processus de Galton-Watson. Ann. Inst. H. Poincaré Probab. Stat. **22**, 199–207 (1986)

204. J. Neveu, Multiplicative martingales for spatial branching processes, in *Seminar on Stochastic Processes 1987*, ed. by E. Çinlar et al. Progress in Probability and Statistics, vol. 15 (Birkhäuser, Boston, 1988), pp. 223–242

205. P. Olofsson, Size-biased branching population measures and the multi-type $x \ln x$ condition. Bernoulli **15**, 1287–1304 (2009)

206. M. Pain, Velocity of the L-branching Brownian motion (2015+). `ArXiv:1510.02683`

207. R. Pemantle, Search cost for a nearly optimal path in a binary tree. Ann. Appl. Probab. **19**, 1273–1291 (2009)
208. Y. Peres, Probability on trees: an introductory climb, in *École d'Été St-Flour 1997*. Lecture Notes in Mathematics, vol. 1717 (Springer, Berlin, 1999), pp. 193–280
209. Y. Peres, O. Zeitouni, A central limit theorem for biased random walks on Galton-Watson trees. Probab. Theory Relat. Fields **140**, 595–629 (2008)
210. J. Peyrière, Turbulence et dimension de Hausdorff. C. R. Acad. Sci. Paris Sér. A **278**, 567–569 (1974)
211. D. Piau, Théorème central limite fonctionnel pour une marche au hasard en environnement aléatoire. Ann. Probab. **26**, 1016–1040 (1998)
212. J.W. Pitman, Some conditional expectations and identities for Bessel processes related to the maximal drawdown of the Brownian meander (2015+, in preparation)
213. Y.X. Ren, T. Yang, Limit theorem for derivative martingale at criticality w.r.t. branching Brownian motion. Stat. Probab. Lett. **81**, 195–200 (2011)
214. P. Révész, *Random Walk in Random and Non-random Environments*, 3rd edn. (World Scientific, Singapore, 2013)
215. M.I. Roberts, A simple path to asymptotics for the frontier of a branching Brownian motion. Ann. Probab. **41**, 3518–3541 (2013)
216. M.I. Roberts, Fine asymptotics for the consistent maximal displacement of branching Brownian motion. Electron. J. Probab. **20**, 1–26 (2015). Paper No. 28
217. R. Rhodes, V. Vargas, Gaussian multiplicative chaos and applications: a review. Probab. Surv. **11**, 315–392 (2014)
218. E. Seneta, On recent theorems concerning the supercritical Galton–Watson process. Ann. Math. Stat. **39**, 2098–2102 (1968)
219. E. Seneta, On the supercritical Galton-Watson process with immigration. Math. Biosci. **7**, 9–14 (1970)
220. B.A. Sevast'yanov, Branching stochastic processes for particles diffusing in a bounded domain with absorbing boundaries. Theory Probab. Appl. **3**, 111–126 (1958)
221. Z. Shi, Random walks and trees. ESAIM Proc. **31**, 1–39 (2011)
222. D. Simon, B. Derrida, Quasi-stationary regime of a branching random walk in presence of an absorbing wall. J. Stat. Phys. **131**, 203–233 (2008)
223. Ya.G. Sinai, The limiting behavior of a one-dimensional random walk in a random medium. Theory Probab. Appl. **27**, 256–268 (1982)
224. F. Solomon, Random walks in a random environment. Ann. Probab. **3**, 1–31 (1975)
225. E. Subag, O. Zeitouni, Freezing and decorated Poisson point processes. Commun. Math. Phys. **337**, 55–92 (2015)
226. H. Tanaka, Time reversal of random walks in one-dimension. Tokyo J. Math. **12**, 159–174 (1989)
227. B. Tóth, No more than three favourite sites for simple random walk. Ann. Probab. **29**, 484–503 (2001)
228. W. van Saarloos, Front propagation into unstable states. Phys. Rep. **386**, 29–222 (2003)
229. V.A. Vatutin, V. Wachtel, Local probabilities for random walks conditioned to stay positive. Probab. Theory Relat. Fields **143**, 177–217 (2009)
230. B. Virág, On the speed of random walks on graphs. Ann. Probab. **28**, 379–394 (2000)
231. S. Watanabe, Limit theorem for a class of branching processes, in *Markov Processes and Potential Theory. Proceedings of a Symposium in Mathematics Research Center*, Madison, WI, 1967 (Wiley, New York, 1967), pp. 205–232
232. E.C. Waymire, S.C. Williams, A cascade decomposition theory with applications to Markov and exchangeable cascades. Trans. Am. Math. Soc. **348**, 585–632 (1996)
233. C. Webb, Exact asymptotics of the freezing transitions of a logarithmically correlated random energy model. J. Stat. Phys. **145**, 1595–1619 (2011)
234. O. Zeitouni, Random walks in random environment, in *École d'Été St-Flour 2001*. Lecture Notes in Mathematics, vol. 1837 (Springer, Berlin, 2004), pp. 193–312
235. O. Zeitouni, Branching random walks and Gaussian fields (2012+). Lecture notes available at: http://www.wisdom.weizmann.ac.il/~zeitouni/pdf/notesBRW.pdf

LECTURE NOTES IN MATHEMATICS  Springer

Editors in Chief: J.-M. Morel, B. Teissier;

Editorial Policy

1. Lecture Notes aim to report new developments in all areas of mathematics and their applications – quickly, informally and at a high level. Mathematical texts analysing new developments in modelling and numerical simulation are welcome.

   Manuscripts should be reasonably self-contained and rounded off. Thus they may, and often will, present not only results of the author but also related work by other people. They may be based on specialised lecture courses. Furthermore, the manuscripts should provide sufficient motivation, examples and applications. This clearly distinguishes Lecture Notes from journal articles or technical reports which normally are very concise. Articles intended for a journal but too long to be accepted by most journals, usually do not have this "lecture notes" character. For similar reasons it is unusual for doctoral theses to be accepted for the Lecture Notes series, though habilitation theses may be appropriate.

2. Besides monographs, multi-author manuscripts resulting from SUMMER SCHOOLS or similar INTENSIVE COURSES are welcome, provided their objective was held to present an active mathematical topic to an audience at the beginning or intermediate graduate level (a list of participants should be provided).

   The resulting manuscript should not be just a collection of course notes, but should require advance planning and coordination among the main lecturers. The subject matter should dictate the structure of the book. This structure should be motivated and explained in a scientific introduction, and the notation, references, index and formulation of results should be, if possible, unified by the editors. Each contribution should have an abstract and an introduction referring to the other contributions. In other words, more preparatory work must go into a multi-authored volume than simply assembling a disparate collection of papers, communicated at the event.

3. Manuscripts should be submitted either online at www.editorialmanager.com/lnm to Springer's mathematics editorial in Heidelberg, or electronically to one of the series editors. Authors should be aware that incomplete or insufficiently close-to-final manuscripts almost always result in longer refereeing times and nevertheless unclear referees' recommendations, making further refereeing of a final draft necessary. The strict minimum amount of material that will be considered should include a detailed outline describing the planned contents of each chapter, a bibliography and several sample chapters. Parallel submission of a manuscript to another publisher while under consideration for LNM is not acceptable and can lead to rejection.

4. In general, **monographs** will be sent out to at least 2 external referees for evaluation.

   A final decision to publish can be made only on the basis of the complete manuscript, however a refereeing process leading to a preliminary decision can be based on a pre-final or incomplete manuscript.

   Volume Editors of **multi-author works** are expected to arrange for the refereeing, to the usual scientific standards, of the individual contributions. If the resulting reports can be forwarded to the LNM Editorial Board, this is very helpful. If no reports are forwarded or if other questions remain unclear in respect of homogeneity etc, the series editors may wish to consult external referees for an overall evaluation of the volume.

5. Manuscripts should in general be submitted in English. Final manuscripts should contain at least 100 pages of mathematical text and should always include

   - a table of contents;
   - an informative introduction, with adequate motivation and perhaps some historical remarks: it should be accessible to a reader not intimately familiar with the topic treated;
   - a subject index: as a rule this is genuinely helpful for the reader.
   - For evaluation purposes, manuscripts should be submitted as pdf files.

6. Careful preparation of the manuscripts will help keep production time short besides ensuring satisfactory appearance of the finished book in print and online. After acceptance of the manuscript authors will be asked to prepare the final LaTeX source files (see LaTeX templates online: https://www.springer.com/gb/authors-editors/book-authors-editors/manuscriptpreparation/5636) plus the corresponding pdf- or zipped ps-file. The LaTeX source files are essential for producing the full-text online version of the book, see http://link.springer.com/bookseries/304 for the existing online volumes of LNM). The technical production of a Lecture Notes volume takes approximately 12 weeks. Additional instructions, if necessary, are available on request from lnm@springer.com.

7. Authors receive a total of 30 free copies of their volume and free access to their book on SpringerLink, but no royalties. They are entitled to a discount of 33.3 % on the price of Springer books purchased for their personal use, if ordering directly from Springer.

8. Commitment to publish is made by a *Publishing Agreement*; contributing authors of multiauthor books are requested to sign a *Consent to Publish form*. Springer-Verlag registers the copyright for each volume. Authors are free to reuse material contained in their LNM volumes in later publications: a brief written (or e-mail) request for formal permission is sufficient.

Addresses:
Professor Jean-Michel Morel, CMLA, École Normale Supérieure de Cachan, France
E-mail: moreljeanmichel@gmail.com

Professor Bernard Teissier, Equipe Géométrie et Dynamique,
Institut de Mathématiques de Jussieu – Paris Rive Gauche, Paris, France
E-mail: bernard.teissier@imj-prg.fr

Springer: Ute McCrory, Mathematics, Heidelberg, Germany,
E-mail: lnm@springer.com